AF668868

THERMAL ENGINEERING II

AUTHOR

Dr. Parvesh Antil

Assistant Professor, Department of Basic Engineering
College of Agricultural Engineering & Technology,
CCS HAU, Hisar 125004, Haryana, India

REVIEWER

Dr. Pankaj Kumar Gupta

Associate Professor, Department of Mechanical Engineering
School of Studies of Engineering and Technology,
Guru Ghasidas Vishwavidyalaya,
Bilaspur 495009, Chhattisgarh, India.

KHANNA BOOK PUBLISHING CO. (P) Ltd.
4C/4344, Ansari Road, Darya Ganj,
New Delhi, India

All India Council for Technical Education
Nelson Mandela Marg, Vasant Kunj,
New Delhi, India

Dear Readers,

To prevent the piracy, this book is secured with HIGH SECURITY HOLOGRAM on the front title cover. In case you don't find the hologram on the front cover title, please write us to at contact@khannabooks.com or whatsapp us at +91-99109 09320 and avail special gift voucher for yourself.

Specimen of Hologram on front Cover title:

Moreover, there is a SPECIAL DISCOUNT COUPON for you with EVERY HOLOGRAM.

How to avail this SPECIAL DISCOUNT:

Step 1: Scratch the hologram

Step 2: Under the scratch area, your "coupon code" is available

Step 3: Logon to www.khannabooks.com

Step 4: Use your "coupon code" in the shopping cart and get your copy at a special discount

Step 5: Enjoy your reading!

ISBN: 978-93-5538-983-1

Book Code: DIP479EN

Thermal Engineering II

by Dr. Parvesh Antil

First Edition: 2026

Originally Published by:

All India Council for Technical Education, Ministry of Education, GOI

Exclusively Published & Distributed by:

Khanna Book Publishing Co. (P) Ltd.

4C/4344, Ansari Road, Daryaganj,
New Delhi (INDIA)

Visit us at: **www.khannabooks.com**

Write us at: **contact@khannabooks.com**

To view complete list of books, Please scan the QR Code: KPH

Printed in Bharat.

Copyright © Reserved

No part of this publication may be reproduced, stored in a retrieval system or transmitted, in any form or by any means, electronic, mechanical, photocopying, recording or otherwise without prior permission of the publisher.

This book is sold subject to the condition that it shall not, by way of trade, be lent, re-sold, hired out or otherwise disposed of without the publisher's consent, in any form of binding or cover other than that in which it is published.

Disclaimer: The website links provided by the author in this book are placed for informational, educational & reference purpose only. The Publisher do not endorse these website links or the views of the speaker/ content of the said weblinks. In case of any dispute, all legal matters to be settled under Delhi Jurisdiction only.

प्रो. टी. जी. सीताराम
अध्यक्ष
Prof. T. G. Sitharam
Chairman

सत्यमेव जयते

आज़ादी का अमृत महोत्सव

अखिल भारतीय तकनीकी शिक्षा परिषद्
(भारत सरकार का एक सांविधिक निकाय)
(शिक्षा मंत्रालय, भारत सरकार)
नेल्सन मंडेला मार्ग, वसंत कुंज, नई दिल्ली–110070
दूरभाष : 011–26131498
ई–मेल : chairman@aicte-india.org

ALL INDIA COUNCIL FOR TECHNICAL EDUCATION
(A STATUTORY BODY OF THE GOVT. OF INDIA)
(Ministry of Education, Govt. of India)
Nelson Mandela Marg, Vasant Kunj, New Delhi-110070
Phone : 011-26131498
E-mail : chairman@aicte-india.org

FOREWORD

Engineers are the backbone of any modern society. They are the ones responsible for the marvels as well as the improved quality of life across the world. Engineers have driven humanity towards greater heights in a more evolved and unprecedented manner.

The All India Council for Technical Education (AICTE), have spared no efforts towards the strengthening of the technical education in the country. AICTE is always committed towards promoting quality Technical Education to make India a modern developed nation emphasizing on the overall welfare of mankind.

An array of initiatives has been taken by AICTE in last decade which have been accelerated now by the National Education Policy (NEP) 2020. The implementation of NEP under the visionary leadership of Hon'ble Prime Minister of India envisages the provision for education in regional languages to all, thereby ensuring that every graduate becomes competent enough and is in a position to contribute towards the national growth and development through innovation & entrepreneurship.

One of the spheres where AICTE had been relentlessly working since past couple of years is providing high quality original technical contents at Under Graduate & Diploma level prepared and translated by eminent educators in various Indian languages to its aspirants. For students pursuing 2nd year of their Engineering education, AICTE has identified 87 books, which shall be translated into 12 Indian languages - Hindi, Tamil, Gujarati, Odia, Bengali, Kannada, Urdu, Punjabi, Telugu, Marathi, Assamese & Malayalam. In addition to the English medium, books in different Indian Languages are going to support the students to understand the concepts in their respective mother tongue.

On behalf of AICTE, I express sincere gratitude to all distinguished authors, reviewers and translators from the renowned institutions of high repute for their admirable contribution in a record span of time.

AICTE is confident that these outcomes based original contents shall help aspirants to master the subject with comprehension and greater ease.

(Prof. T. G. Sitharam)

ACKNOWLEDGEMENT

I am deeply grateful to the authorities of AICTE, with special acknowledgment to Prof. T. G. Sitharam, Chairman; Dr. Abhay Jere, Vice Chairman and Prof. Rajive Kumar, Member-Secretary and Dr. Sunil Luthra, Director, Training and Learning Bureau. Their collective vision and planning were instrumental in the publication of the book on **Thermal Engineering II.** I extend my sincere appreciation to Prof. Pankaj Kumar Gupta, Associate Professor, Department of Mechanical Engineering, School of Studies of Engineering and Technology, Guru Ghasidas Vishwavidyalaya, Bilaspur, Chhattisgarh, India. As the reviewer of this book, Dr. Pankaj Gupta has made invaluable contributions, ensuring the content is not only student-friendly but also aesthetically pleasing and well-organized.

Special acknowledgment goes to my wife, Khushi, who provided invaluable support during the completion of the book. I wish to thank my parents and family members, whose nurturing and support have been my pillars through the significant events of my life. Lastly, I extend my profound thanks to my beloved kids, Raavi and Shaurya, for their understanding and patience, excusing my work during weekends and family time. This book is the result of a variety of suggestions from AICTE members, experts, and authors who shared their opinions and thoughts on how to further develop engineering education in our country. Acknowledgements are due to the contributors and different workers in this field, whose published books, review articles, papers, photographs, footnotes, references, and other valuable information enriched me at the time of writing the book.

Dr. Parvesh Antil

PREFACE

The book "Thermal Engineering II" was made possible by the knowledge that I acquired while teaching thermal engineering courses at the Department of Basic Engineering, COAE&T, CCS HAU Hisar, Haryana. This book is specifically designed to teach the fundamental principles of thermal engineering to diploma students. The course's primary objectives are as follows:

- ✓ The aim is to provide a thorough grasp and profound insight into all essential components associated with gas turbines, steam turbines, nozzles, and steam generators.
- ✓ To study, analyze, and evaluate the properties of steam, entropy, and enthalpy and comprehend the Mollier chart for different processes.
- ✓ The goal is to comprehend the principles and functions of different turbines and boilers.

This book's first unit discusses gas turbines and their classification. The unit's final segment delves into the details of jet propulsion, Ram jet engines, turbojet engines, and rocket engines. The second unit delves into the properties of steam, encompassing its formation process and other essential definitions. We also present a detailed discussion on enthalpy, entropy, dryness fraction, the Mollier chart, and calorimeters. Unit three concentrates on steam boilers, covering their functions and classifications. The detailed description of Cochran, Babcock, and Wilcox boilers, Lamont, and Benson boilers, along with various boiler accessories, mountings, and draught systems, is included in this unit. The fourth unit provides a detailed explanation of nozzles and how the flow velocity through them affects the heat drop. The effect of friction and methods of calculation of cross-sectional areas at throat and exit for maximum discharge are discussed to make it simple for the reader. The fifth unit contains comprehensive information on steam turbine classification and operation. We have used the velocity diagrams and derivations to discuss the working of impulse and reaction turbines. To help students learn, each chapter ends with multiple-choice questions, short and long answer questions, and exercise problems. To help students understand, each chapter ends with books and online lectures from relevant courses.

I hope this book inspires diploma students to understand and discuss thermal engineering concepts. I welcome any comments and recommendations to enhance this book in future versions.

Dr. Parvesh Antil

OUTCOME BASED EDUCATION

For the implementation of an outcome based education the first requirement is to develop an outcome based curriculum and incorporate an outcome based assessment in the education system. By going through outcome based assessments, evaluators will be able to evaluate whether the students have achieved the outlined standard, specific and measurable outcomes. With the proper incorporation of outcome based education there will be a definite commitment to achieve a minimum standard for all learners without giving up at any level. At the end of the programme running with the aid of outcome based education, a student will be able to arrive at the following outcomes:

Programme Outcomes (POs) are statements that describe what students are expected to know and be able to do upon graduating from the program. These relate to the skills, knowledge, analytical ability attitude and behaviour that students acquire through the program. The POs essentially indicate what the students can do from subject-wise knowledge acquired by them during the program. As such, POs define the professional profile of an engineering diploma graduate.

National Board of Accreditation (NBA) has defined the following seven POs for an Engineering diploma graduate:

PO1. Basic and Discipline specific knowledge: Apply knowledge of basic mathematics, science and engineering fundamentals and engineering specialization to solve the engineering problems.

PO2. Problem analysis: Identify and analyses well-defined engineering problems using codified standard methods.

PO3. Design/ development of solutions: Design solutions for well-defined technical problems and assist with the design of systems components or processes to meet specified needs.

PO4. Engineering Tools, Experimentation and Testing: Apply modern engineering tools and appropriate technique to conduct standard tests and measurements.

PO5. Engineering practices for society, sustainability and environment: Apply appropriate technology in context of society, sustainability, environment and ethical practices.

PO6. Project Management: Use engineering management principles individually, as a team member or a leader to manage projects and effectively communicate about well-defined engineering activities.

PO7. Life-long learning: Ability to analyse individual needs and engage in updating in the context of technological changes.

COURSE OUTCOMES

By the end of the course the students are expected to learn:

CO-1: Various types of turbines and jet propulsion system.

CO-2: Formation of steam and various properties on Mollier chart.

CO-3: Classify steam generators and discussion on types of boilers, their accessories and mountings.

CO-4: Flow, velocity and discharge of steam through nozzles.

CO-5: About impulse and reaction turbines through their velocity diagrams. Also, learn about the governing of steam turbines through schematic diagrams.

Mapping of Course Outcomes with Programme Outcomes to be done according to the matrix given below:

Unit Outcomes	EXPECTED MAPPING WITH PROGRAMME OUTCOMES (1- Weak Correlation; 2- Medium Correlation; 3- Strong Correlation)						
	PO-1	**PO-2**	**PO-3**	**PO-4**	**PO-5**	**PO-6**	**PO-7**
CO-01	3	2	3	1	3	3	3
CO-02	3	3	3	2	3	2	3
CO-03	3	3	3	3	3	3	3
CO-04	3	2	3	3	3	2	3
CO-05	3	2	3	3	3	2	3

GUIDELINES FOR TEACHERS

To implement Outcome Based Education (OBE) knowledge level and skill set of the students should be enhanced. Teachers should take a major responsibility for the proper implementation of OBE. Some of the responsibilities (not limited to) for the teachers in OBE system may be as follows:

- Within reasonable constraint, they should manoeuvre time to the best advantage of all students.
- They should assess the students only upon certain defined criterion without considering any other potential ineligibility to discriminate them.
- They should try to grow the learning abilities of the students to a certain level before they leave the institute.
- They should try to ensure that all the students are equipped with the quality knowledge as well as competence after they finish their education.
- They should always encourage the students to develop their ultimate performance capabilities.
- They should facilitate and encourage group work and team work to consolidate newer approach.
- They should follow Blooms taxonomy in every part of the assessment.

Bloom's Taxonomy

Level	Teacher should check	Student should be able to	Possible mode of assessment
Create	Student's ability to create	Design or Create	Mini project
Evaluate	Student's ability to justify	Argue or Defend	Assignment
Analyse	Student's ability to distinguish	Differentiate or Distinguish	Project/Lab Methodology
Apply	Student's ability to use information	Operate or Demonstrate	Technical Presentation/ Demonstration
Understand	Student's ability to explain ideas	Explain or Classify	Presentation/Seminar
Remember	Student's ability to recall	Define or Recall	Quiz

GUIDELINES FOR STUDENTS

Students should take equal responsibility for implementing the OBE. Some of the responsibilities (not limited to) for the students in OBE system are as follows:

- Students should be well aware of each UO before the start of a unit in each and every course.
- Students should be well aware of each CO before the start of the course.
- Students should be well aware of each PO before the start of the programme.
- Students should think critically and reasonably with proper reflection and action.
- Learning of the students should be connected and integrated with practical and real life consequences.
- Students should be well aware of their competency at every level of OBE.

ABBREVIATION & SYMBOLS

Abbreviation	Full Form	Abbreviation	Full Form
C_p	Specific heat at constant pressure	U	Internal energy
C_v	Specific heat at constant volume	L	Latent heat
I.C Engines	Internal combustion engines	L_i	Internal latent heat
atm	atmospheric	L_e	External latent heat
P	Pressure	T_{sup}	Superheated Temperature
T	Temperature	W_e	Equivalent evaporation
s	entropy	FD	Forced draught
H	Enthalpy	C-D	Convergent-Divergent
h	Specific enthalpy	A	Area
η_{th}	Thermal efficiency	P_i	Inlet pressure
q	heat	K	Nozzle efficiency
r_p	Pressure ratio	V_s	Velocity of steam
C	Constant	k	Blade velocity
Q_S	Heat supplied	$\dot{m}$	Mass flow rate
Q_R	Heat rejected	η_{stage}	Stage efficiency
T_{cold}	Temperature of cold reservoir	η_b	Blade efficiency
T_{hot}	Temperature of hot reservoir	R	Degree of reaction
°C	Degree Celsius	α	Alpha
P_c	Critical pressure	β	Betta
T_c	Critical temperature	Φ	Phi
kJ	Kilojoule	θ	Theta
kg	Kilogram	γ	Gamma(Adiabatic index)
x	Dryness fraction	F	Reheat factor
m_s	Mass of steam		

LIST OF FIGURES

CONTENTS

1 Gas Turbines

UNIT SPECIFICS

This unit presents information related to the following topics:

- Basics of gas turbines, their working and description with p-v and T-S diagrams.
- Various types/layouts of gas turbines and comparison with IC engines and steam turbines.
- Principle of jet propulsion, fuel used for the propulsion and various applications.
- Working of rocket engine and comparison with jet propulsion.

This unit provides readers with a fundamental understanding of gas turbines. Understanding the fundamental principles of these turbines is crucial for comprehending their operation under varying temperature and pressure conditions. The author provides specific information on p-v and T-S diagrams to enhance the readers' understanding of the process. This category also encompasses the classification and evaluation of gas turbines, comparing them to reciprocating internal combustion engines and steam turbines. The course provides an overview of the general layout of both open cycle constant pressure gas turbines and closed cycle gas turbines. This course also includes comprehensive information on jet propulsion, aiming to give readers a concise and efficient understanding of the subject. This unit aims to provide readers with a comprehensive understanding of the fundamental principles of rocket propulsion, the functioning of a rocket engine, and its diverse range of applications. The unit concludes with a concise review of the key concepts. Additionally, the unit includes a significant portion of multiple-choice and descriptive questions. The unit includes a list of readings and sources for students to utilize for practice. It is important to mention that QR codes have been included in several parts to give more information on various areas of interest. The readers can scan these codes to obtain pertinent and corroborating information. The unit emphasizes the inclusion of video resources and QR codes to provide more information on numerous areas of interest. These resources can be accessed and viewed on mobile phones by browsing or scanning QR codes.

RATIONALE

The gas turbines are an important part of industrial growth. These turbines play a crucial role in electricity production since they convert mechanical energy into electrical energy via the usage of a generator, which is responsible for the majority of power generation. The primary use of gas turbines is in aviation, where they serve as the main source of power for jet propulsion. Due to the importance of this application and the variety of current jet engines, a detailed discussion of the issue will be provided in a distinct part of the unit. This discussion will focus on the use of gas turbines in the production of electric power and in certain industrial operations. Additionally, it will examine their function in marine, locomotive, and vehicle propulsion. These days, jet fighters travel through the stratosphere at a speed exceeding that of sound. Modern jet airliners have the capability to reach higher altitudes, achieve greater speeds, and cover longer distances than ever before. Jet propulsion is a method that accelerates jets and missiles to their intended destinations. Furthermore, rockets propel Earth satellites into orbit. While jet propulsion is often associated with flying, it may also be used in hydraulic jet propulsion systems for tiny, high-speed boats and recreational vessels. This includes detailed explanation on the gas turbines and jet/rocket propulsion system.

PRE-REQUISITES

Before reading this unit, the students are advised to revisit the following:
Physics: Thermodynamics (Class XII)
Basic Mechanical Engineering (MEPC102)

UNIT OUTCOMES

After studying this unit students will be able to:

U1-O1: Understand the basic of gas turbines and operating cycle.

U1-O2: Learn about the effectiveness of gas turbines as compared to steam turbines and IC engines.

U1-O3: Understand the layout of gas turbines based on operating cycle.

U1-O4: Learn about the jet propulsion and turbojet engines.

U1-O5: Understand the working of Ram jet and rocket engines.

Unit 1 Outcomes	EXPECTED MAPPING WITH COURSE OUTCOMES (1- Weak Correlation; 2- Medium Correlation; 3- Strong Correlation)				
	CO-1	CO-2	CO-3	CO-4	CO-5
U1-O1	3	1	-	-	-
U1-O2	3	-	-	-	-
U1-O3	3	-	-	-	-
U1-O4	3	-	-	-	-
U1-O5	3	-	-	-	-

1.0 Air standard cycles

Air standard cycles are thermodynamics concept used to evaluate internal combustion engine efficiency. These cycles simplify engine thermodynamics. Engineers test gas turbine and internal combustion engine efficiency using air standard cycles like the Otto, Diesel, and Brayton cycles. These cycles are standardized models that allow engineers to assess engine efficacy and efficiency. These cycles may help engineers understand engine thermodynamics. They can make informed design decisions and improve engine performance.

1.1 Air-standard Brayton cycle

Figure 1.1 displays a schematic representation of an air-standard gas turbine. The arrows on this diagram represent the principal energy transfer directions. An air-standard analysis assumes that heat transfer from an external source to the working fluid causes the temperature increase during the combustion process. The working fluid is assumed to be air, which behaves as an ideal gas. Under the air-standard idealizations, the compressor would intake air at state 1 from the surrounding environment and then release it back to the surroundings at state 4, at a temperature higher than the surrounding temperature. Following its interaction with the environment, every unit of discharged air will ultimately revert back to its original condition, similar to the air that enters the compressor. Therefore, we may consider the air flowing through the gas turbine components as going through a thermodynamic cycle. The air goes through a simplified representation of its stages in this cycle by perception of the air coming out of the turbine as being returned to the air coming in at the compressor after passing through a heat exchanger that rejects heat to the environment. Further conceptualization leads to a process known as the air standard Brayton cycle.

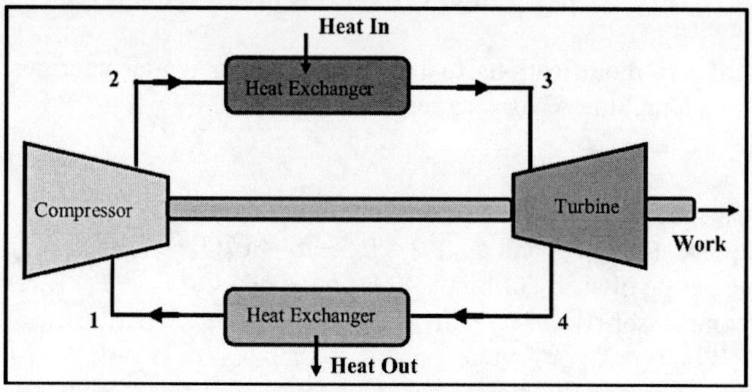

Fig. 1.1 Air-standard gas turbine

The air is compressed isentropically in the compressor, heated at constant pressure in the heat exchanger, expanded isentropically in the turbine, and then cooled back to its starting condition before returning to the compressor. The assumptions behind the Brayton cycle are as follows:

- The working fluid is air, which is treated as an ideal gas.
- The combustion process occurring in the combustion chamber is replaced by the continuous supply of heat at a constant pressure from an external source.

- The exhaust process is substituted by a constant pressure heat rejection in a heat exchanger.

The Brayton cycle is composed of four processes that are internally reversible.

Process 1-2: Isentropic compression of air in the compressor.

Process 2-3: Includes the addition of heat to air at a constant pressure.

Process 3-4: Involve the isentropic expansion of air within the turbine.

Process 4-1: Involves the constant pressure heat rejection of air.

➤ **Description with p-v and T-S diagrams**

Fig. 1.2 represents the p–V and T–s diagrams for an ideal Brayton cycle. The Brayton cycle's four operations are executed in a consistent sequence. Consequently, they are analyzed as constant flow processes.

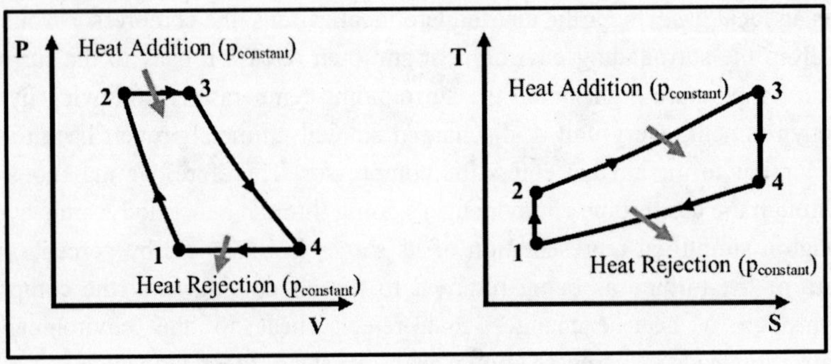

Fig. 1.2 p-V and T-s Diagrams for Brayton Cycle

In the absence of any modifications to the kinetic and potential energies, the steady-flow energy equation on a unit mass basis, as represented by

q-w = $heat_{rejected}$- $heat_{added}$

Δh= $heat_{rejected}$- $heat_{added}$

Let's assume that specific heat of air is constant, then

$q_{addition}= h_3 - h_2 = C_p (T_3 - T_2)$ and $q_{rejected} = h_4 - h_1 = C_p(T_4 - T_1)$ ....(1.1)

Now, work done per kg of air by turbine $(W_T) = h_3 - h_4 = C_p (T_3 - T_4)$ ...... (1.2)

Input work on compressor $(W_C) = h_2 - h_1 = C_p (T_2 - T_1)$ ....(1.3)

Net work done (W_{net})* = Σq or Σw; i.e. $heat_{added}$ - $heat_{rejected}$ or $W_T - W_C$

(*as per first law of thermodynamics, net work done in a thermodynamic cycle is the net work done is equal to the net head added)

Now the thermal efficiency will be $\eta_{Brayton} = \frac{w_{net}}{q_{added}} = 1 - \frac{q_{rejected}}{q_{added}} = 1 - \frac{C_p (T_4 - T_1)}{C_p (T_3 - T_2)}$

$$\eta_{Brayton} = 1 - \frac{T_1 (T_4/T_1 - 1)}{T_2 (T_3/T_2 - 1)} \quad \text{.......................} \ (1.4)$$

The process 1-2 and process 3-4 are isentropic processes as their entropy is constant.

Therefore, $\frac{T_2}{T_1} = \left(\frac{p_2}{p_1}\right)^{\frac{\gamma-1}{\gamma}} = (r_p)^{\frac{\gamma-1}{\gamma}}$ .... (1.5)

and $\frac{T_3}{T_4} = \left(\frac{p_3}{p_4}\right)^{\frac{\gamma-1}{\gamma}} = \left(\frac{p_2}{p_1}\right)^{\frac{\gamma-1}{\gamma}} = (r_p)^{\frac{\gamma-1}{\gamma}}$ ....(1.6)

As we know, p_2=p_3 and p_4=p_1; we get $\frac{T_2}{T_1} = \frac{T_3}{T_4}$ or $\frac{T_4}{T_1} = \frac{T_3}{T_2}$

Using these values, the equation becomes $\eta_{Brayton} = 1 - \frac{1}{r_p^{(\gamma-1)/\gamma}}$ .... (1.7)

Here r_p= pressure ratio and $\gamma = \frac{C_p}{C_v}$

1.2 Gas turbines

A gas turbine is a rotating internal combustion engine consisting of a compressor, a combustion chamber, and a turbine. The gas turbine utilizes a steady stream of gas as the working medium to convert a part of thermal energy into mechanical energy. The engine produces gas by means of fuel combustion. The blades of the turbine wheel are subjected to streams of gas that are expelled by stationary nozzles. The jet's propulsion force induces the rotation of the shaft. Fig. 1.3 illustrates a gas turbine operating on a single cycle. The axial compressor intakes ambient air and raises its pressure and temperature through compression. The combustion chamber receives the compressed air at a high temperature and injects fuel in the form of either gas or liquid spray. Combustion is facilitated by a burner. The combustion by products produced are utilized to propel a turbine, so creating electrical power. The turbine releases the exhaust gasses into the atmosphere. The turbine power is mostly used by the compressor and ancillary devices, with the remaining amount being the net plant output. The resulting energy can be used for power generation, aircraft propulsion, or other applications.

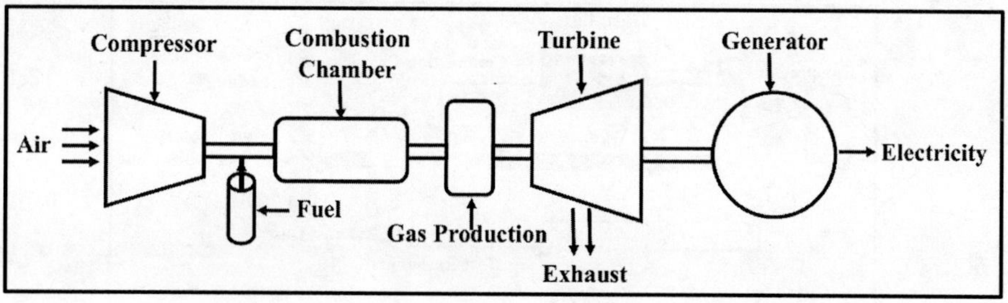

Fig. 1.3 Schematic diagram of Simple Gas Turbine

1.3 Gas turbines Classification

The gas turbines are rotary engines and can be categorized as follows:

a) Open-cycle Gas Turbines

It is the simplest form of the gas turbine cycle which includes a compressor and rotor. The compressor and rotor rotates together. When the air gets compressed then the compressor transfers the compressed air to the combustion chamber. Energy is supplied by injecting fuel into the air stream within the combustion chamber. The high-temperature gases discharged from the combustion chamber undergo expansion as they pass through the turbine and are released into the atmosphere. In order for the unit to generate net-work output, the turbine must provide a greater amount of gross work output than is required to operate the compressor and compensate for mechanical losses in the drive. The fan expels the combustion byproducts into the atmosphere once they are no longer usable. The operating fluids, namely gasoline and air, require regular replacement due to their use during operation. Therefore, the cycle transforms into an open cycle, as seen in Figure 1.4.

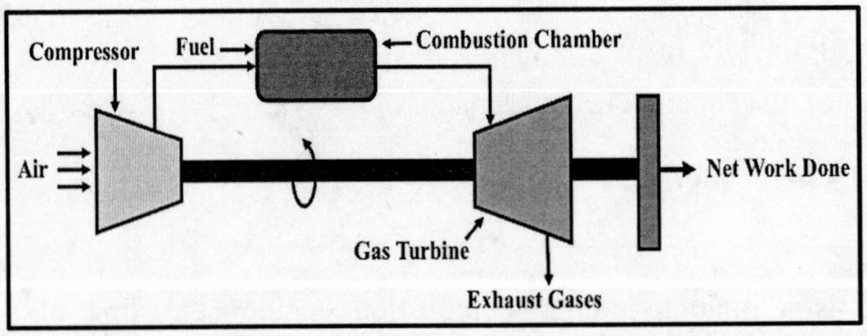

Fig. 1.4 Open cycle gas turbine

b) Closed-cycle Gas Turbine

In a closed-cycle gas turbine, the working fluid, typically air or a suitable material, exits the compressor and undergoes heating at a constant pressure by an external heat source. Afterwards, the air that is experiencing high pressure and high temperature is directed towards the turbine. Following its passage through the turbine, the fluid undergoes a cooling process facilitated by an external cooling agent, which restores it to its initial temperature prior to being returned to the compressor. This closed-loop methodology enables the continuous utilization of fluids while limiting any alterations in their state. The schematic diagram of closed cycle gas turbine is illustrated in fig. 1.5.

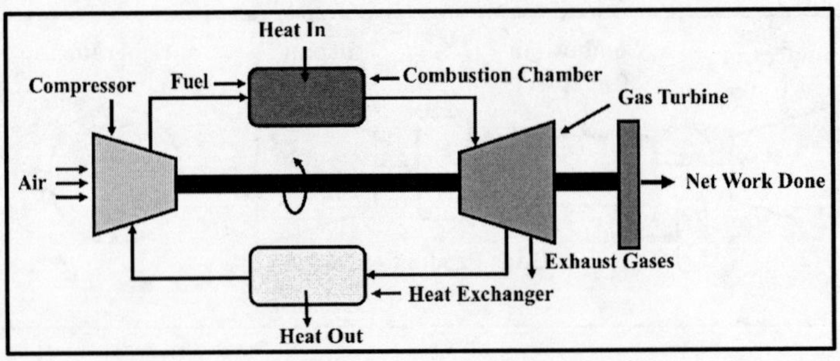

Fig. 1.5 Closed cycle gas turbine

1.4 Comparison of gas turbine with reciprocating I.C. engines and steam turbines

The gas turbines have the following merits and demerits over reciprocating IC engines.

- ***Merits***

1. Mechanical Efficiency
   The gas turbine's mechanical efficiency is 95%, higher than IC engines, due to its rotating nature.
2. Flywheel
   A gas turbine does not need a flywheel since its shaft torque is constant and uniform, but an IC engine cannot function smoothly without one.
3. Weight
   The gas turbine weighs less per kW than the IC engine.
4. Power
   Gas turbines provide higher power per kg of air than IC engines.
5. Speed
   Gas turbines are faster than IC engines.
6. Fuel

Gas turbines employ cheaper fuel like paraffin, oil dregs, or powdered coal, whereas IC engines require special-grade lubricants to minimize knocking.

7. Pollution level
   Gas turbine exhaust is cleaner than IC engine exhaust because combustion is complete.
8. Operative systems
   Gas turbines have simpler ignition and lubricating systems than IC engines.

➢ ***Demerits***

1. Gas turbine power plants have 15-20% lower thermal efficiency than IC engines (25-30%).
2. Gas turbines are difficult to start than IC engines.
3. Gas turbine blade cooling requires special equipment.
4. Gas turbine blade production is expensive and complicated.
5. Gas turbines create five times more exhaust gas than IC engines for the same power output.

1.5 Comparison between a gas turbine and a steam turbine

The comparison between a gas turbine and steam turbine on the basis of key characteristics is explained below:

- Working fluid

Steam turbines use steam as a working fluid which is the cheapest medium and can be created using readily accessible resources. Gas turbines burn expensive fuel efficiently in their combustion chambers.

- Floor area

A steam power plant has a boiler, turbine, condensing plant, cooling tower, and water supply. All steam and power plant equipment is big and takes up space. A gas turbine plant has just three parts: compressor, combustion chamber, and turbine. Also compact is the combustion chamber. Gas turbine engines are lightweight and utilized in aviation.

- Speed

For speed, gas turbines are superior than steam turbines.

- Thermal Efficiency

The continuous improvement in the technology has made the steam turbines more thermally efficient than gas turbine.

- Combustion Chamber

The steam turbines have external combustion whereas gas turbine has internal combustion chamber.

- Working fluid pressure

The working fluid pressure ratio in gas turbine is lower as compared to steam turbine.

- Cooling System

Gas turbine plants need no cooling except blade cooling, but steam power plant condensers require a lot of cooling water.

➢ **Applications and limitations of gas turbines**

Gas turbines find applications in the following areas:

- Power plants utilise gas turbines to produce energy.
- Gas turbines power jet engines used in aircraft.

- Several cruise ships and naval vessels are propelled by gas turbines.
- Mechanical systems and industrial equipment are driven by them.
- Oil drilling and processing use gas turbines.
- For pipeline-based natural gas transportation.
- Various compressors and pumps are powered by gas turbines.
- Used in backup generators.
- Gas turbines are used to power some tanks and automobiles.

➢ ***Limitations of Gas Turbines***

- The thermal efficiency of gas turbines is lower than that of internal combustion engines.
- The closed gas turbine cycle cannot be used in aviation engines since it needs cooling water to function.
- Gas turbines have a lower maximum combustion pressure than internal combustion engines.
- Due to its bulkiness and lack of a heater, the constant volume explosion type gas turbine is not suitable for use in aircraft.
- The only reason that the maximum gas temperature is restricted to around 750°C is the potential for blade material failure. The blade is made of an alloy of nickel and cobalt.

1.6 General lay-out of Open cycle constant pressure gas turbine

In this particular type, the fuel undergoes combustion at a consistent pressure, utilising the Joule cycle as its operating principle. The turbine is of the reaction type and operates using oil fuel. It is equipped with an axial flow rotary air compressor that is connected to the rotor shaft. The heat is provided to the working fluid under a consistent pressure. A basic open cycle constant pressure gas turbine (as shown in figure 1.4) is made up of a compressor, combustion chamber, and turbine. The air is drawn in from the atmosphere and compressed to a combustion pressure of 1 to 4 atm by the compressor. It is then compelled into the combustion chamber. A portion of the air is utilised for combustion purposes, while the rest is directed through the annular space between the combustion chamber wall and the burner jacket. The air absorbs heat from the burner jacket and combines with the by-products of the combustion chamber and the burner jacket. This increases the temperature and volume of the air. By utilising a significant amount of air beyond what is needed for combustion, the temperature of the mixture is kept at a level that is safe for the metal of the rotor blades. Additionally, it helps to regulate the temperature of the burner.

The gases under intense pressure and temperature enter the turbine and pass through the blade rings. As they flow, they gradually expand, transforming the pressure energy into kinetic energy. This energy is then absorbed by the rotor. After exiting the turbine, the exhausted gases flow out through the exhaust. The speed of the turbine is determined by regulating the fuel oil supply and the safety valve, which diverts a portion of the mixture. When the pressure gets too high, the turbine uses its power to drive the generator. However, a significant portion of the power produced by the rotor is used to drive the compressor. The turbine is activated by the electric starter motor. In the closed cycle constant pressure gas turbine, the compressed fluid from the compressor is heated in the heater by an external source at constant pressure. The high pressure and temperature fluid then generates work as it passes through the turbine. The fluid discharged from the turbine is cooled back to its initial temperature in the cooler, utilising an external cooling source, prior to entering the compressor. Therefore, the working fluid is constantly circulated within the closed cycle. This type of turbine is no longer in use.

- **p-V and T-s diagrams and working**

Open cycle gas turbines are made up of three main components: a compressor, a power turbine unit and a combustor. A gas turbine in an open cycle operates by taking in ambient air and compressing it using either centrifugal or axial flow compressors. The compressor efficiently processes the atmospheric air by effectively compressing it through a series of meticulously designed compressor stages. Air and fuel are combined in the combustion chamber after being injected. Intense, scorching gases are utilised to power the turbine. Once the mixture is ignited, it produces velocity gas. As the high-speed gas flows through the turbine blades, it causes the connected shaft to rotate, which in turn powers the generator's rotor section. The power generated by spinning the turbine shaft has the ability to fuel a wide range of industrial machinery and can also be harnessed to produce electrical energy. Exhaust from turbines is expelled into the air. Because the exhaust gases are not recirculated but instead released into the atmosphere, the system is considered an open cycle. The turbine utilises a portion of its power to operate the compressor and other machinery, while the rest is harnessed for the production of electricity. The compressor receives a constant supply of fresh air, while the turbine expels its gases into the atmosphere, ensuring a continuous replacement of the working medium.

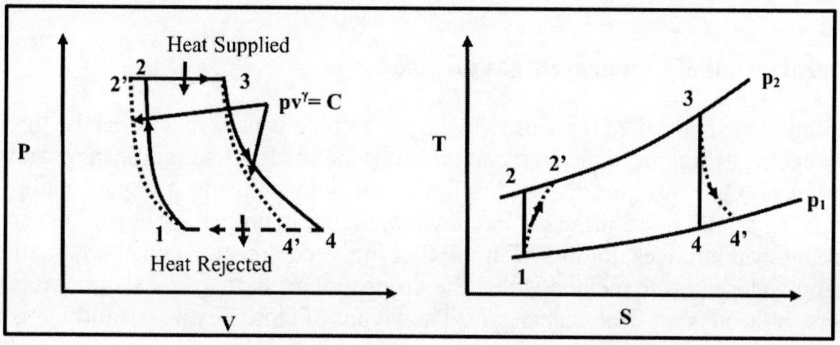

Fig. 1.6 p-V and T-s Diagram of open cycle gas turbine

The stages of operations as shown in fig. 1.6 are described below:

Stage 1 *to* 2' (Isentropic compression): The atmospheric air enters the compressor at pressure (p_1) and temperature (T_1). After that, the atmospheric air compressed isentropically to temperature T_2 and pressure p_2.

Stage 2' *to* 3 (Constant pressure heat addition) The heat is added in this stage at constant pressure which rises the temperature of the air to T_3.

Stage 3 *to* 4' (Isentropic expansion) The air having high temperature and pressure then expands at constant entropy in the turbine from pressure p_2 to p_4.

Stage 4' *to* 1 (Constant pressure heat rejection) The heat generated from the burnt gases after combustion are exhausted at constant pressure (p).

The area 1-2'-3-4' of the *P-V* diagram shows the net-work output in an ideal case.

- **Thermal Efficiency**

The thermal efficiency of the open cycle gas turbine can be expressed as ratio of net work output to the heat supplied and can be expressed mathematically as

$$\eta_{th} = \frac{W_{net}}{q_{supplied}}$$

During stage 2'-3, heat is added at constant pressure, then the equation becomes

$$\eta_{th} = \frac{W_{turbine} - W_{compressor}}{(h_3 - h_2)} \quad \text{.................................} (1.8)$$

$$\text{Here, } W_{turbine} = h_3 - h_4 = C_p(T_3 - T_4)$$

$$W_{compressor} = h_2 - h_1 = C_p(T_2 - T_1)$$

$$q_{supplied} = h_3 - h_2 = C_p(T_3 - T_2)$$

Therefore,

$$\eta_{th} = \frac{(h_3 - h_4) - (h_2 - h_1)}{(h_3 - h_2)} \quad \text{......................................} (1.9)$$

Also,

$$\eta_{th} = \frac{(T_3 - T_4) - (T_2 - T_1)}{(T_3 - T_2)} \quad \text{.....................................} (1.10)$$

1.7 General layout of Closed cycle gas turbine

An alternative method called the closed-cycle gas turbine is used to address the limitations of the open cycle gas turbine. The degradation of turbine blades is a significant disadvantage in an open cycle. One way to address this limitation is by utilising a higher quality working medium, such as air or various gases like helium, argon, hydrogen, or neon. This ensures that the working medium does not mix with the fuel in the combustion chamber. Another benefit of utilising a closed cycle method is that the dissipation of heat from exhaust gases occurs in re-coolers, re-heaters, or heat exchangers. The layout of closed cycle gas turbine is shown in figure 1.5.

- **p-V and T-s diagrams and working**

The working principle of the closed-cycle gas turbine is derived from the Brayton cycle or Joule's cycle. In this particular gas turbine design, the compressor is responsible for uniformly compressing the gas, which is then directed into the heating chamber. This turbine is best suited for the rotor type compressor. A external supply is used for heating up the compressed air before it flows over the turbine blades. As the gas flows over the turbine blades, it undergoes expansion and is then directed into the cooling chamber to be cooled down. The gas is cooled by circulating water at a constant pressure, bringing it back to its initial temperature. The gas is then sent back into the compressor and the process is repeated. In this turbine, the gas is continuously circulated throughout the system. Using a different working fluid/medium in the turbine would lead to an increase in both the complexity of the system and its cost. This could potentially result in some challenges and may prove to be quite challenging to address.

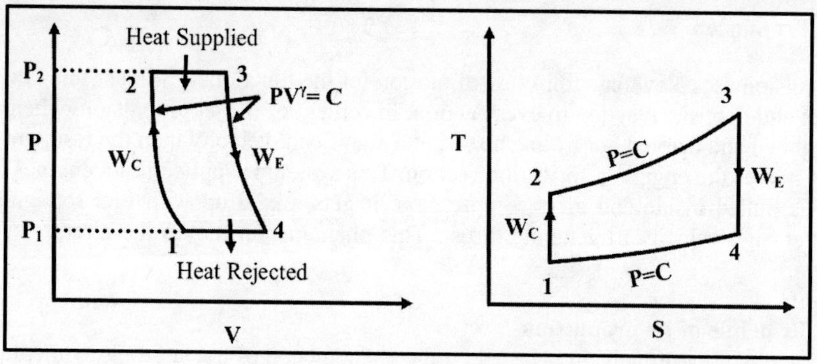

Fig. 1.7 p-V and T-s diagrams

Process 1-2: Isentropic compression; Process 2-3: Heat addition at constant process; Process 3-4: Isentropic expansion; Process 4-1: Heat rejection at constant process

Process 1-2: The air is compressed isentropically from a lower to a greater pressure, causing a temperature rise. No heat is transferred.

Process 2-3: The system suffers a rise in volume from V_2 to V_3 and temperature from T_2 to T_3, but the pressure stays constant at P_2. This results in a heat flow into the system. The formula for calculating heat received is mc_p (T_3–T_2).

Process 3-4: The air undergoes an isentropic expansion, resulting in a decrease in pressure from p_2 to p_1 and a corresponding drop in temperature from T_3 to T_4. There is no transfer of heat.

Process 4-1: The heat is released from the system as the volume decreases from V_4 to V_1 and the temperature drops from T_4 to T_1, while the pressure remains constant at p_1. The formula for calculating heat rejected is mc_p (T_4-T_1).

The air standard efficiency can be termed as ratio of work done to the heat supplied.

$\eta_{air\text{-}standard}$ = work done/heat supplied

$\eta_{air\text{-}standard}$ = (heat supplied-heat rejected)/heat supplied

$$= mc_p(T_3 - T_2) - mc_p(T_4 - T_1)/mc_p(T_3 - T_2)$$

$$= 1 - \frac{(T_4 - T_1)}{(T_3 - T_2)} \quad \text{.................................... (1.11)}$$

As process 1-2 and process 3-4 are isentropic processes,

Therefore

$$\frac{T_2}{T_1} = \left(\frac{p_2}{p_1}\right)^{\frac{\gamma-1}{\gamma}} \quad \text{........................ (1.12)}$$

$$\frac{T_2}{T_1} = (r_p)^{\frac{\gamma-1}{\gamma}}; \; T_2 = T_1(r_p)^{\frac{\gamma-1}{\gamma}}$$

Similarly,

$$\frac{T_3}{T_4} = \left(\frac{p_2}{p_1}\right)^{\frac{\gamma-1}{\gamma}}$$

$$\frac{T_3}{T_4} = (r_p)^{\frac{\gamma-1}{\gamma}}; T_3 = T_4(r_p)^{\frac{\gamma-1}{\gamma}}$$

Now, $\eta_{air\text{-}standard} = 1 - \dfrac{T_4 - T_1}{T_4(r_p)^{\frac{\gamma-1}{\gamma}} - T_1(r_p)^{\frac{\gamma-1}{\gamma}}}$ ....(1.13)

$$\eta_{air\text{-}standard} = 1 - \frac{1}{(r_p)^{\frac{\gamma-1}{\gamma}}} \quad \text{.................................... (1.14)}$$

1.8 Jet Propulsion

Jet propulsion uses Newton's third law of motion for the functioning. As per this law, there is an equal and opposite reaction to every action. In context of the jet propulsion, when a part of gas is ignited and exerted out from a nozzle at a significant velocity then the resultant force of this gas moves the engine in forward direction. This system is applied in jet engines in which the air is pulled inside and after compression, it gets mixed up with fuel for ignition and ejected at high velocity to generate thrust. This phenomenon moves the aircraft in forward direction.

➢ **Principle of jet propulsion**

The principle of jet propulsion is based on the phenomenon of transmitting momentum to the fluid mass so that resulting reaction produces propulsive force. This phenomenon can be employed by using a nozzle for expanding the highly pressurised hot gas. This expansion creates a strong jet of heated gases in the atmosphere, which reacts to propel the opposite way. The open cycle gas turbine is best for jet propulsion.

➢ **Fuels used for jet propulsion**

Modern aviation heavily depends on jet propulsion to achieve high speed and long distance travel. Jet fuels are used to effectively power and operate modern aircraft. Jet fuel is a primary source of fuel for jet propulsion. Aviation necessitates the use of gasoline as a fuel source to propel aircraft and enable them to achieve take off. Aviation fuels exhibit a range of distinct qualities and are used for various purposes.

These aviation fuels are customized to particular aircraft engines and operating circumstances. Jet aircraft and turboprop engines utilize jet fuel, or Aviation Turbine Fuel. Jet fuels include kerosene-based Jet A and Jet A-1 and naphtha-based Jet B. Its parallels to diesel are evident. Planes, which have changed how we travel, now use numerous fuels. Many aircraft utilize jet fuel, but others operate on gasoline, a popular car fuel. The recommended airplane fuel is aviation gasoline (AvGas). It is modified as per aviation needs. AvGas powers piston-engine airplanes and other aircraft. It powers these aircraft, enabling them to do their jobs. The formulation of AvGas makes it ideal for aircraft fuels for jet engines. Aviation fuel is varied. Jet aircraft and turboprop engines use Aviation Turbine Fuel (ATF), also known as Jet A and Jet A-1 (kerosene) and Jet B (naphtha).

➢ **Applications of jet propulsion**

The various applications of jet propulsion are following:

- Jet propulsion is essential for missiles to reach their objectives quickly. Guided missiles employ this propulsion mechanism to attack their targets at great speeds. Due to jet propulsion, missiles can move faster and more accurately in wartime.
- Rockets are essential for launching Earth satellites. Satellites are launched into orbit using strong propulsion systems.
- By combining rocket power, scientists and engineers can position satellites in their orbits despite Earth's gravity.
- Jet propulsion has uses beyond flying. Small, high-speed boats and leisure ships use hydraulic jet propulsion. Jet propulsion is flexible and adaptive; despite being associated with aviation.

1.9 Working of a turbojet engine

Fig.1.8 displays a schematic of a turbojet engine along with the T-s diagram (fig. 1.9) of the ideal turbojet cycle. The air pressure increases slightly as it slows down in the diffuse. The compressor compresses the air. The fuel is combined with air in the combustion chamber and undergoes combustion at a consistent pressure. The combustion gases undergo partial expansion in the turbine, generating sufficient power to operate the compressor and other associated machinery. At last, the gases undergo expansion in a nozzle, reaching the ambient pressure and exiting the engine with great speed.

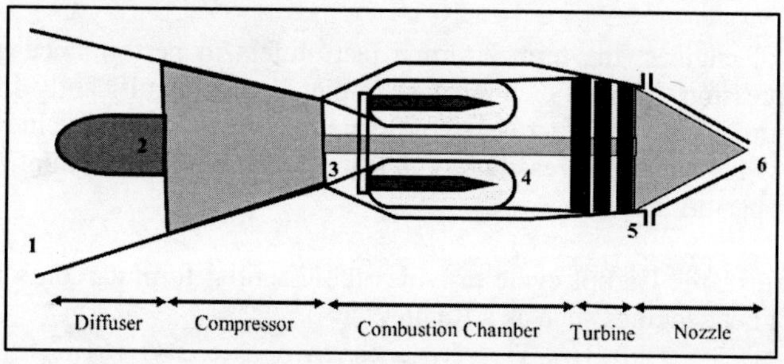

Fig. 1.8 Schematic diagram of Turbojet Engine

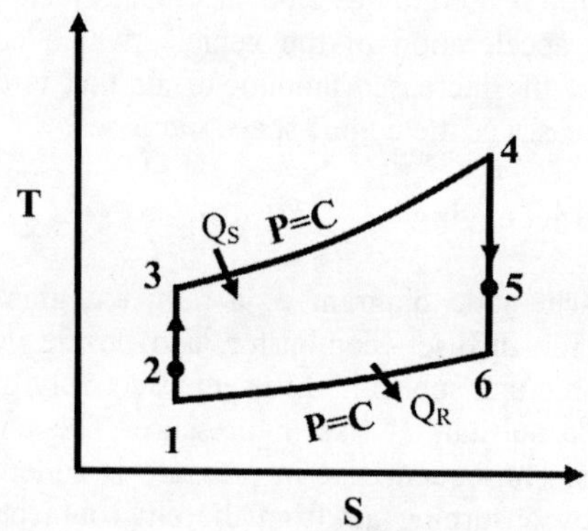

Fig 1.9. T-s Diagram of turbojet engine

The various stages in a turbojet engine are following:

Stage 1-2: Inlet

Stage 2-3: Compression

Stage 3-4: Combustion

Stage 4-5: Turbine

Stage 5-6: Nozzle

The system includes a diffuser at the entrance that reduces the speed of the air, converting some of its kinetic energy into pressure. This process is known as ram compression. The air is further compressed to a pressure of 3 to 4 bar in a rotary compressor, typically of the axial flow type. The combustion chamber (C.C.), into which fuel is fed, receives compressed air. The fuel burns at a quite constant pressure, which raises the temperature quickly. The heated gases travel on to the gas turbine, where they expand somewhat. The compressor, fuel pump and other supporting equipment may be run on the electricity produced. By use of Newton's third law of motion, the aircraft is propelled forward by a strong jet created by the gas turbine's exhaust gases, which are expelled via a nozzle at a greater pressure than the surrounding air.

1.10 Principle of Ram effect

In the context of engines, the term "Ram Effect" refers to performance gains because of intake air compression. The efficiency of a Ramjet is also significantly influenced by the compression ratio. Therefore, increasing pressure recovery may be achieved by reducing diffusion process losses and increasing combustion process efficiency, both of which would enhance Ramjet performance.

Within this context, the Carnot cycle provides the essential formula controlling the thermal efficiency (η) of heat engines such as a Ramjet:

$$\eta = 1 - T_{cold}/T_{hot} \quad \text{.............................} (1.15)$$

T_{cold}= Temperature of cold reservoir and T_{hot} = Temperature of hot reservoir
It is essential to have an adequate knowledge of the Ram Effect while discussing the concept of efficiency. This expression is used to describe the enhancement in engine performance that occurs as a result of the acceleration of the vehicle, which causes the intake air to be compressed. As a result of the increased amount of air that is driven into the engine, the efficiency of a Ramjet increases as the engine speed increases.

- **Working of a Ram jet engine**

Figure 1.10 depicts the schematic diagram of a ramjet engine. The system comprises a supersonic diffuser, subsonic diffuser, combustor, and nozzle designated as a, b, c and d respectively. Both supersonic and subsonic diffusers effectively transform the kinetic energy of the incoming air into a substantial rise in pressure. The ram effect is the process of converting energy, and the consequent rise in pressure is called ram pressure. The engine raises its static pressure by absorbing air from the environment at a high speed and then slowing down in the supersonic diffuser. After that, the air is forced even further into the subsonic diffuser, where it rises in temperature and pressure to the levels needed for ignition. Then the air flows smoothly into the combustion chamber, where it combines with the fuel and any air that is yet unburned. Potential energy of the gases is transformed into motion energy by their channelling into the nozzle. The ramjet needs forward push, and that comes from the high-velocity gasses that exit the nozzle. The main benefits of ramjet engines are their low weight, absence of moving components, and fuel flexibility. Its need of a launching mechanism and inability to start on its own from a fixed position are disadvantages, too. Due to its remarkable propulsion at high velocities, it has gained popularity as a preferred option for high-speed military aircraft and missiles. Subsonic ramjets are used in guided munitions, in conjunction with turbojets or rockets, to provide the required initial rotational force.

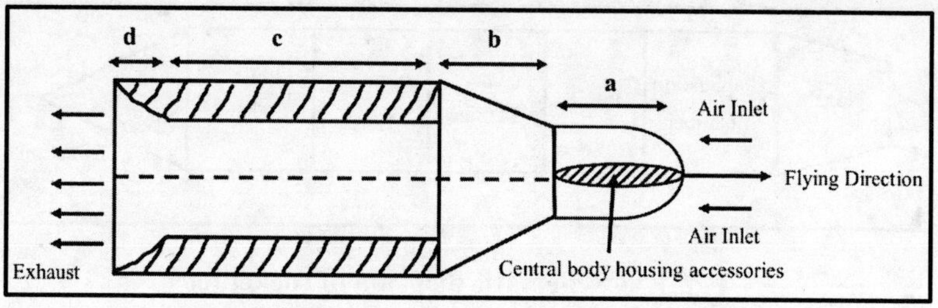

Fig. 1.10 Schematic diagram of ramjet engine

1.11 Principle of Rocket propulsion

The rocket propulsion system is accountable for powering the rocket, lifting it off the ground, and forcing it into the upper atmosphere. Rockets, unlike jets, are supplied with independent propellant. Although rocketry has a history spanning thousands of years, its current evolution just started in the last few centuries. In recent times, rockets are traveling to very long distances with the help of the propellant which can be in solid or liquid form. Apart of sending machines and rovers to the Mars, the scientist also trying to send humans to the Mars with the help of these advance space crafts and propellants. The basic fundamental of rocket propulsion is based on Newton's third law of motion which states that every action has equal and opposite reaction. When it comes to rockets, the "action" refers to the propellant being shot downward, and the "reaction" that consequently occurs is the rocket being blasted upward. It is the mass of the propellant as well as its acceleration that determines the amount of force that is applied.

➢ **Working principle of a rocket engine**

The propellant body of a rocket engine contains both the fuel and the oxidizer. The rocket is capable of operating in a vacuum and is the sole apparatus capable of space travel. Rockets can be categorised as either single-stage or multi-stage, and they are composed of one or more rocket engines. These rockets can either have solid propellant or liquid propellant. Figure 1.11 is a schematic representation of a liquid propellant rocket engine that utilises liquid oxygen and purified petrol. The system comprises a fuel tank, oxidizer tank, gas turbine, pumps, combustion chamber, and nozzle. Pumps are utilised to deliver propellants to the nozzle with elevated pressure in order to achieve more thrust. The operation of these pumps relies on a gas turbine, which receives the high pressure and high temperature gases produced by the fuel and oxidizer. An ideal propellant should possess several desirable features, including a high calorific value, high density, high stability, ease of handling and storage, non-corrosiveness, and a high boiling point at low pressure. Rockets have significant uses in several fields such as long-range artillery, deadly weaponry, signalling and fireworks, jet-assisted take-off, satellites, space ships, and research. They are also capable of autonomous operation.

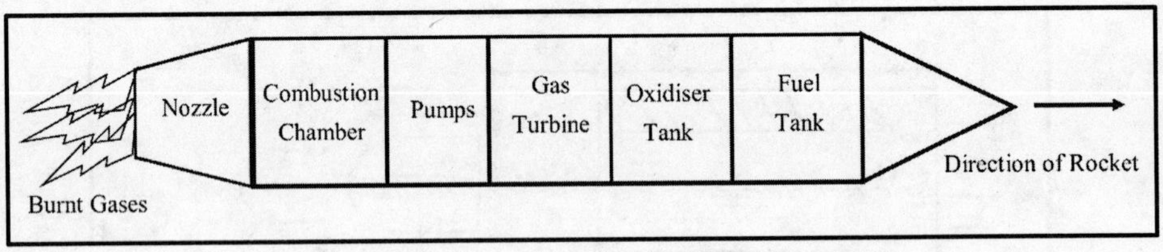

Fig. 1.11 Schematic diagram of rocket jet

➢ Applications of rocket propulsion

The common applications of rocket propulsion are following:

- The most common application of rocket propulsion is in jet engines.
- High speed cars also uses rocket propulsion for their speed.
- It is used in satellite launching process.
- It is also useful for exploring space through space vehicles.
- The rocket propulsion helps in inter planetary travels to explore life and other useful information.

➢ Comparison of jet and rocket propulsions

There are following few characteristics on which we can compare jet and rocket propulsion:

1. ***Peripheral Source of Starting***

The rocket propulsion starts with the oxidisation of fuel but jet propulsion need external source for starting compressor and turbine.

2. ***Supply of Oxidant***

The jet propulsion doesn't carry oxidant supply but rocket propulsion has in built oxidiser tank.

3. ***Compressor and Turbine***

The compressor and turbine are installed in the jet propulsion but there is no requirement of turbine and compressor in the rocket propulsion.

4. ***Utility in outer space***

The rocket propulsion carries oxidiser tank which helps in travel to outer space but jet propulsion cannot operate in outer space as it requires atmospheric air for burning of fuel and it does not carry any oxidiser tank.

5. ***Propulsive Thrust***

Both jet and rocket moves forward with the reaction of engine generated by high velocity gases produced after burning of fuel. They achieve the desired thrust by common method.

UNIT SUMMARY

Based on the contents covered in this unit you should remember that:

- ❖ Air standard cycles are thermodynamics concept used to evaluate internal combustion engine efficiency.
- ❖ Gas turbine is an internal combustion rotary engine using combustion gases as working fluid.
- ❖ The gas turbines can be broadly classified in to two groups as open cycle gas turbines and closed cycle gas turbines.

- ❖ Ramjet engine uses engine's forward motion to compress the incoming air.
- ❖ The efficiency of a Ramjet increases with the increase in engine speed.
- ❖ In turbojet engine, thrust is produced due to acceleration of hot combustion gases through the exhaust nozzle.
- ❖ The basic difference between jet and rocket propulsion is that the jet propulsion uses atmospheric oxygen and stores fuel, whereas rocket engines use fuel and oxidizer in a propelling body and may operate in vacuum.
- ❖ Rockets can be classified on the basis of motors and propellants.

Exercises

Multiple Choice Questions

1. What is the primary purpose of a gas turbine?

A) To generate electrical power

B) To pump fluids

C) To compress air

D) To store energy

2. Which component of a gas turbine is responsible for compressing the incoming air?

A) Combustor

B) Turbine

C) Compressor

D) Nozzle

3. In a gas turbine, what is the function of the combustor?

A) To mix fuel with air and ignite it

B) To expand the high-pressure air

C) To cool the exhaust gases

D) To filter the incoming air

4. What type of gas turbine cycle is most commonly used in commercial power generation?

A) Otto cycle

B) Brayton cycle

C) Diesel cycle

D) Rankine cycle

5. Which part of the gas turbine extracts energy from the high-temperature, high-pressure gases to perform work?

A) Compressor

B) Turbine

C) Combustor

D) Heat exchanger

6. **What is the typical efficiency range of modern gas turbines?**

    A) 10% - 20%

    B) 20% - 30%

    C) 30% - 40%

    D) 40% - 50%

7. **Which of the following is a major advantage of using a gas turbine compared to a steam turbine?**

    A) Higher operating temperature

    B) Lower noise levels

    C) Lower initial cost

    D) Faster start up and shutdown times

8. **What is the primary material used for the turbine blades in gas turbines to withstand high temperatures?**

    A) Aluminium

    B) Stainless steel

    C) Super alloys

    D) Titanium

9. **In a gas turbine, what happens to the air as it passes through the compressor?**

    A) It is cooled and compressed

    B) It is heated and expanded

    C) It is compressed and heated

    D) It is cooled and expanded

10. **What is the primary difference between a simple cycle gas turbine and a combined cycle gas turbine?**

    A) The type of fuel used

    B) The presence of a heat recovery steam generator (HRSG) in the combined cycle

    C) The number of compressors used

    D) The operating pressure

Answer Key:

1 A	2C	3 A	4 B	5 B
6 D	7 D	8 C	9 C	10 B

Short and Long Answer Type Questions

- What do you mean by gas turbines?
- What are the classifications of gas turbines?

- Describe with neat sketches the working of a simple constant pressure open cycle gas turbine.
- Compare gas turbine with reciprocating I.C. engines and steam turbines.
- Explain construction and working of turbojet.
- Explain construction and working of open cycle constant pressure gas turbine using P-V and T-S diagrams.
- What is the principle of jet propulsion? Discuss the fuels used for jet propulsion.
- What is Ram effect. Explain the working of a Ram jet engine.
- Compare the jet and rocket propulsions in detail.
- What are the advantages and disadvantages of closed cycle over the open cycle?

Know More

✓ In thermodynamic cycles, gas turbines use the Brayton cycle, which involves adiabatic compression, isobaric heat addition, expansion, and rejection. To increase efficiency, modern turbines may utilize variants such as the regenerative Brayton cycle or mixed cycles (which integrate with steam turbines).

✓ Materials for high-performance gas turbines have to withstand high temperatures and stresses. Turbine blades uses single-crystal super alloys for their strength and thermal fatigue resistance.

✓ Variable geometry blades in modern gas turbines strengthen performance under difficult operational conditions. The cooled air from the compressor is carried via diverse internal cooling systems to cool the blades.

✓ Intercooling compressed air before combustion and reheating the expanding gases can improve gas turbine efficiency. The combined cycle power plants through gas and steam turbines may surpass 60% efficiency.

✓ Advanced turbines use lean premixed combustion to minimize NOx emissions by burning completely at lower temperatures.

✓ Gas turbines may generate significant noise, although there are many noise reduction devices used, including improved mufflers and acoustic liners. In addition, improving the blade shape and compressor stages may effectively decrease noise levels.

References and Suggested Readings

➢ Thermal Engineering by B. R. Sarkar (McGraw-Hill Education (India) Pvt Limited).

➢ Thermal Engineering by R.K. Rajput (Laxmi Publications (P) Ltd).

➢ A Textbook of Thermal Engineering by R.S. Khurmi & J.K. Gupta (S. Chand & Company Ltd).

Dynamic QR Code for Further Reading

1. Lectures of the NPTEL Course on "Gas Turbines", by Prof. Ravi Kumar, Department of Mechanical & Industrial Engineering, Indian Institute of Technology Roorkee, India.

2. Lectures of the NPTEL Course on “Introduction to rocket propulsion” by Dr D. P. Mishra, Department of Aerospace Engineering, Indian Institute of Technology Kanpur, India.

2 Properties of Steam

UNIT SPECIFICS

This unit presents information related to the following topics:

- Formation of steam under constant pressure and various uses in industrial applications.
- Key definitions related to liquid and vapour lines, determination of enthalpy, internal energy and entropy.
- Use of Mollier chart for various process like Isochoric, Isobaric, Hyperbolic, Isothermal, Isentropic, Throttling and Polytropic process.
- Types of calorimeters and numerical.

This unit provides readers with a fundamental understanding of steam. The author provides specific information on formation of steam to enhance the readers' understanding of the process. This unit also includes the key definitions which are very useful in understanding the formation and further utilization of steam as per need. The unit includes detailed explanation on the dryness fraction and behaviour of steam at higher temperature. This course also includes comprehensive information on calorimeters and its types, aiming to give readers a concise and efficient understanding of the topic.

The unit concludes with a concise review of the key concepts. Additionally, the unit includes a significant portion of multiple-choice and descriptive questions. The unit includes a list of readings and sources for students to utilize for practice. It is important to mention that QR codes have been included in several parts to give more information on various areas of interest. The readers can scan these codes to obtain pertinent and corroborating information. The unit emphasizes the inclusion of video resources and QR codes to provide more information on numerous areas of interest. These resources can be accessed and viewed on mobile phones by browsing or scanning QR codes.

RATIONALE

Steam formed at a higher pressure has higher temperature and can be made to flow easily through insulated pipes from steam generator to point of use. When water is heated in a boiler to create steam, it absorbs energy. The formed steam can hold five or six times as much potential energy as an equivalent mass of water. This energy can easily be transported and used for manufacturing process where direct or indirect heating is needed. These qualities of steam make itself important aspect for study. The readers will get benefit from this unit by learning the basics of steam formation. As the unit progress, the readers will learn about the various important terms which will be helpful in advancing the subject. The analysis of steam on the Mollier chart will help readers to check the quality of steam in the various applications.

PRE-REQUISITES

Before reading this unit, the students are advised to revisit the following:
Physics: Thermodynamics (Class XII)
Basic Mechanical Engineering (MEPC102)

UNIT OUTCOMES

After studying this unit students will be able to:
U2-O1: Understand the basic of steam formation and uses.
U2-O2: Learn about the various terms used in steam analysis.
U2-O3: Understand the steam quality measuring techniques.
U2-O4: Learn about the Mollier chart for various processes.
U2-O5: Understand the types of steam calorimeter.

Unit 2 Outcomes	**EXPECTED MAPPING WITH COURSE OUTCOMES** (1- Weak Correlation; 2- Medium Correlation; 3- Strong Correlation)				
	CO-1	**CO-2**	**CO-3**	**CO-4**	**CO-5**
U2-O1	2	3	2	-	-
U2-O2	-	3	2	-	-
U2-O3	-	2	-	-	-
U2-O4	2	3	1	-	-
U2-O5	-	3	-	-	-

2.0 Steam

Steam is the term given to microscopic droplets of water floating in air after being converted into a gaseous form known as water vapour (figure 2.1). The steam is released from hot springs, fumaroles, and some types of volcanoes as a result of the heating of subterranean water by volcanic processes. Technological systems, such as those using fossil fuel-burning boilers and nuclear reactors, may also produce steam on a huge scale. When steam is produced, it expands. Each litre of heated water expands to 1600 litres of steam.

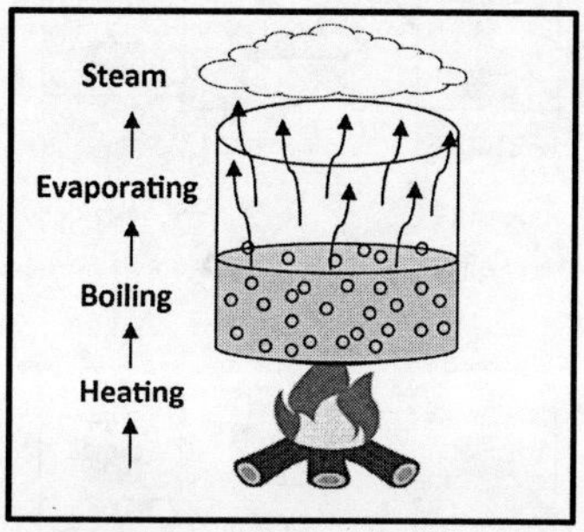

Figure 2.1 Steam generation

The steam is generally divided into two major groups:

- Wet Steam

Wet steam is a mixture of steam and liquid water. It exists at a saturation temperature with more than 5% water content. Due to the presence of water droplets (without changing phase), it is referred to as a two-phase mixture. Machine parts like turbine blades, low-pressure steam pipes, and heat exchangers are susceptible to corrosion caused by wet steam. The steam's ability to transfer heat is reduced when it is wet, which makes sterilisation less effective. Supersaturated steam is another term for wet steam.

- Dry Steam

Dry steam is saturated steam that does not include water particles in suspension. It contains essentially minimal moisture (0.5%) and a very high dryness fraction. The moisture in the steam corrodes and shortens the lifespan of boilers and other heat exchangers. Dry steam is thus preferred in heating applications since it has a superior heat exchange capacity and doesn't lead to corrosion.

2.1 Formation of Steam

The water is converted into steam at a pressure that is generally higher than atmospheric pressure. Suppose we heat up 1 kg of water, which is at absolute 0°C, in a cylinder with a frictionless piston (figure 2.2). Placing a weight (W) on the piston will cause it to heat up at a constant pressure P (stage 1). The water gets warm up as heat is applied to it and keeps rising until it reaches saturation temperature at pressure P.

The heat is applied to raise the water's temperature from 0°C to T°C (stage 2). As the temperature of water continues to rise, heat is used to shift the phase from liquid to vapour. This process continues until all of the water has evaporated into steam, as illustrated in stage 3. The piston is pushed higher, and it can be seen that there is no increase in temperature. The temperature and volume of steam will start to rise if heating is sustained, as illustrated in stage 4. This type of steam is referred to as superheated steam.

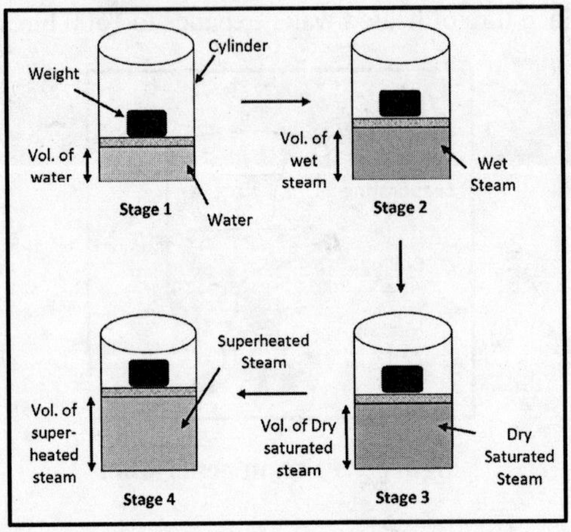

Figure 2.2 Stages of formation of Steam

2.2 Formation of steam under constant pressure

At a temperature of -10°C, or below the freezing point, consider 1 kg of ice and heat it up while maintaining pressure P. The ice absorbs its sensible heat during this process with increase in temperature until it reaches the melting point of ice, which is 0°C. The ice begins to melt by influence of heat. The temperature stays constant at 0°C as it absorbs latent heat of fusion, and it totally transforms into water at 0°C (figure 2.3).

Now, as more heat is added, the temperature of the water begins to rise until it reaches the saturation or boiling point, which corresponds to pressure P. When the temperature reaches the boiling point, it stays steady with subsequent heat additions, and vaporisation occurs.

Despite the continual addition of heat, even after reaching the boiling point, the vaporisation process continues to occur. At the same saturation temperature, the water absorbs its latent heat and transforms into dry, saturated steam. Wet steam, which is actually a combination of steam and water, is the intermediate state between water and dry, saturated steam.

The temperature of this dry saturated steam begins to rise above saturation temperature as more heat is supplied, and it then transforms into superheated steam. Sensible heat is again being absorbed. Degree of superheat is the term used to describe the overall increase in temperature of superheated steam over the saturation temperature. It is important to note that steam's saturation temperature, latent heat, and other parameters stay constant under constant pressure, but they change when pressure fluctuates.

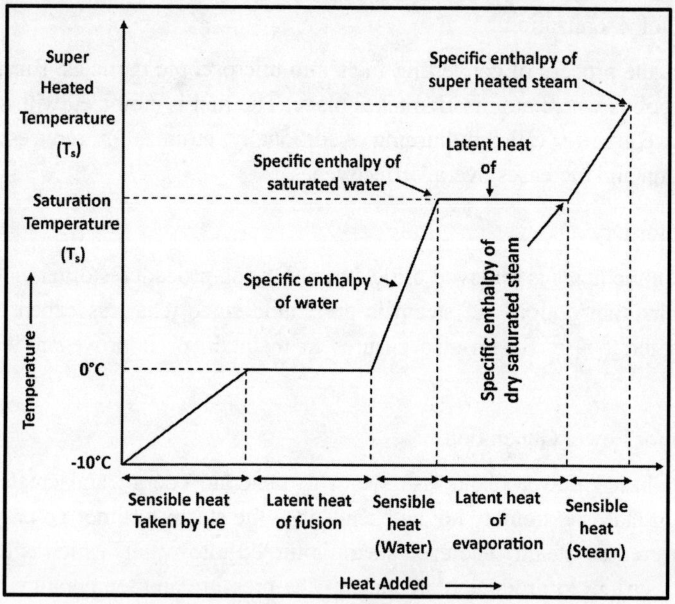

Figure 2.3 Formation of steam under constant pressure

2.3 Industrial use of steam

Steam is used in industries for various purposes. Around the world, the industrial and process sectors depend heavily on steam. Steam boilers have proven to be a crucial component for high-quality production and maximum productivity, especially with the rise in global demand and changes in the manufacturing industry. Following are the areas where steam is used extensively:

❖ Steam for Heating

Steam is widely used for a variety of purposes, including process heating. Steam is the favoured heating medium over alternatives due to a number of features including high efficiency, dependability, cost effectiveness, and ease of use. Steam can be utilised for indirect or direct heating.

✓ Direct Heating

In direct heating, the object that has to be heated is immediately exposed to the steam. To ensure uniform heating and prevent the steam from leaving the steam boiler without heating the items, accuracy is necessary. Examples of major industries that use steam for direct heating include pharmaceuticals and food manufacturing.

✓ Indirect Heating

Steam is used for indirect heating, where there is no direct contact between steam and the object to be heated. With the use of different heating devices, such as heat exchangers, cookers, jacketed containers, etc., indirect heating is accomplished.

❖ Steam for Atomization:

Atomization is the process of converting fuels into microscopic particles. Burners in a steam boiler are the points where atomization takes place. The fuel is atomized with steam to create a greater surface area for efficient burning. Additionally, atomization reduces the amount of smoke formation and increases overall effectiveness.

❖ Steam for Drying

The moisture in the items is removed using steam. Drying procedures often include the use of hot air. Hot air is an option, but steam is preferable since it is less expensive, simpler to handle, and safer to use. Steam also ensures to maintain or improve product quality and finishing.

❖ Steam for Power Generation

Rankine cycle-based power plants use steam to generate energy and create electricity. It produces steam that is extremely hot and sends it to the steam turbine. To produce energy, a turbine is powered by steam. The used steam is turned into water, which is then re-injected into the boiler system to produce more steam. The pressure and temperature of steam at the inlet and output of the turbine have a direct impact on the effectiveness of the power plants, making superheated steam the optimum fuel for power generation.

❖ Steam for Sterilization and Disinfection

Industries involved in processing and manufacturing adhere to set standards to guarantee a clean and secure workplace. It is essential to sterilise and disinfect the products, machinery, and processing facilities. Industries including pharmaceuticals, food processing, chemicals, distilleries, etc. make sure that process plants and equipment are properly sterilised with high-temperature steam to retain the quality of the goods.

2.4 Key Definitions used in Steam Analysis

The fundamental concept that underlies many different thermal relations is steam. The important definitions or concepts in those thermal relations which help in understanding the various thermal changes are listed below.

2.4.1 Saturated liquid line and saturated vapour line

According to variations in the substance's volume, pressure, and temperature, a pure substance will change its state. The saturated liquid line is the boundary separating the compressed or sub-cooled liquid from the saturated mixture (Figure 2.4). The saturated liquid line indicates a substance's transition from a liquid to a vapour. The substance is completely liquid below the saturated liquid line and partially vaporised above it until enough heat is applied to completely vaporise it. Due to vaporisation, a substance takes up a greater specific volume above its saturated liquid line. Saturated vapor is that vapor who, at a given temperature and pressure, will condense into a liquid quickly enough to maintain a constant pressure when its volume is compressed at a constant temperature. The saturated vapour line is a curve where all of the mixture has totally vaporised.

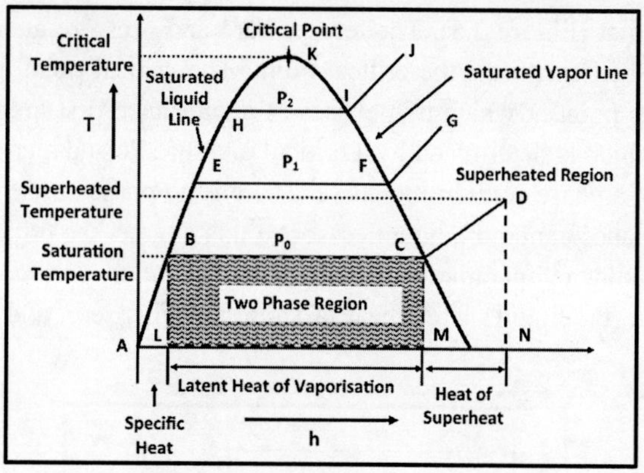

Figure 2.4 T-h diagram showing saturated liquid and saturated vapour lines

2.4.2 Liquid region, vapour region, wet region and superheat region

On the p-h plot, enthalpy is represented by the horizontal ordinates, whereas pressure is represented by the vertical ordinates (Figure 2.5).

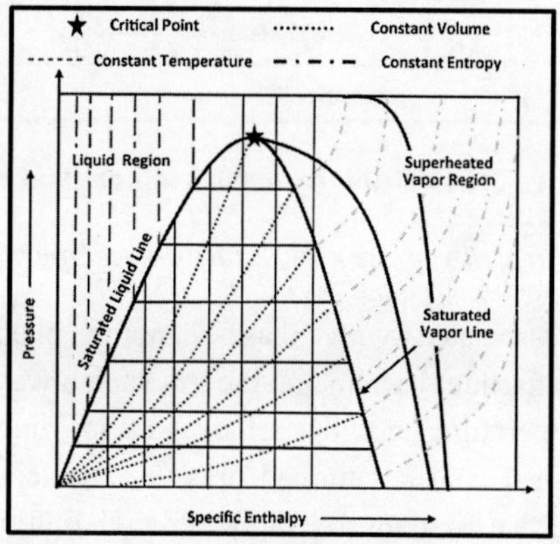

Figure 2.5 p-h diagram showing various regions (phases) of water

The critical point is where the saturated liquid and saturated vapour lines intersect. The region of subcooled liquid is located to the left of the saturated liquid line. The wet region refers to the region that is between the saturated liquid and vapour lines. A substance exists as superheated vapour in the area to the right of the saturated vapour line and at temperatures higher than the critical point temperature. The temperature and pressure are no longer dependent characteristics since the superheated region is a single-phase region (containing just the vapour phase), making it easier to utilise them as the two independent values in the steam tables.

2.4.3 Critical point

The saturation line becomes a point if the pressure of liquid water is steadily raised to the point where the saturated liquid state and the saturated vapour state are identical. This is called the critical point (Figure 2.6). The temperature and pressure at which liquid and gas cannot be distinguished constitute the critical point. The critical point, or critical state, in a phase diagram is the point at which two phases of a substance first merge into one another. The critical point, which is determined by a critical pressure Pc and a critical temperature Tc, is the culmination of a phase equilibrium curve. There is no phase border at this point in time. Since the pure substance cannot be liquefied above this degree, the critical point denotes the maximum boiling point temperature and maximum saturation pressure for a single-phase system. At this point, the liquid's latent heat of vaporisation is zero, and it immediately turns into vapour.

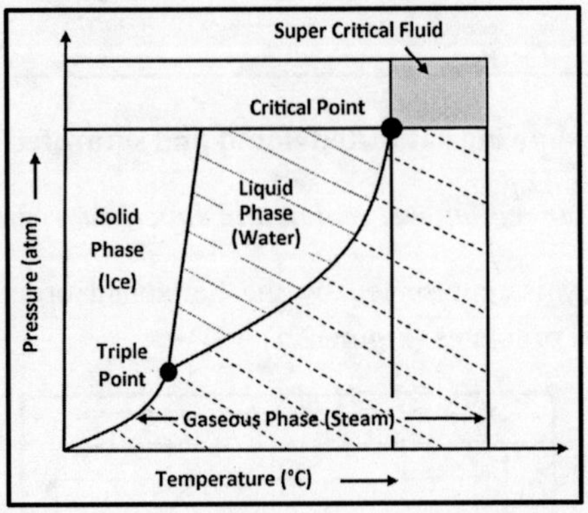

Figure 2.6 P-T diagram showing critical point

2.4.4 Saturated liquid, Saturated vapour and Saturation temperature

The saturated word is associated with phase-changing phenomena (figure 2.7). The temperature and pressure of a liquid are maintained in such a way that, by reducing pressure slightly at a constant temperature, it will start evaporating liquid. In general, a liquid is regarded as saturated when it is at a saturated pressure and temperature. By referring to a saturated liquid as a liquid that is about to evaporate, its definition is clear. Water maintains its liquid state at standard conditions of 20 °C and 1 bar of pressure. If we maintain the same pressure and raise the temperature by about 100 °C, water rapidly enters the condition of vaporisation with only a small change in temperature.

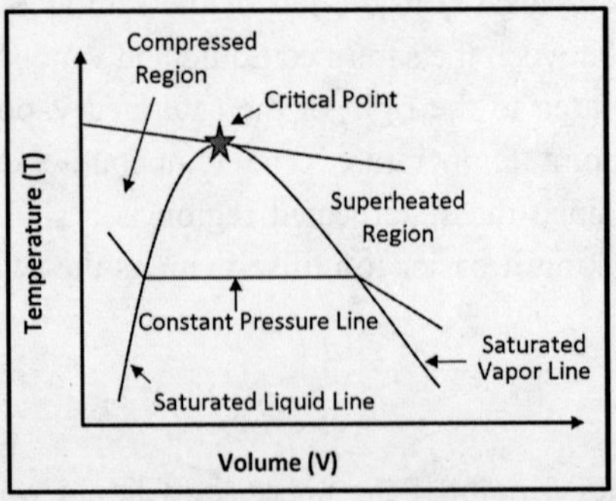

Figure 2.7 T-*v* diagram saturated stages

A vapor whose temperature and pressure are such that any compression of its volume at constant temperature causes it to condense to liquid at a rate sufficient to maintain a constant pressure is called saturated vapor. The maximum amount of vapour that a given volume of air can store at a specific temperature is known as saturated vapour. For instance, the term "saturated water vapour" refers to the most water vapour that a given volume of air can contain at a particular temperature. Vapor that fails to reach its maximum limit is known as unsaturated vapour. However, it does not exist in a saturated vapour state. Dry saturated vapour is also devoid of any liquid particles. In other words, when all liquid particles have been vaporised, dry vapour results. As a result, the condensation of liquid particles in the vapour can be caused by any rise in vapour pressure or drop in vapour temperature.

The temperature at which a liquid transforms into its vapour phase is known as the saturation temperature. Thermal energy has been stated to have saturated the liquid. A phase transition is brought on by any increase in thermal energy.

2.4.5 Sensible heat

Sensible heat is literally the heat that can be felt. It is the energy moving from one system to another that changes the temperature, rather than changing its phase. For example, it warms water rather than melting ice. In other words, the amount of heat that a body can withstand without having an impact on its molecular structure is known as sensible heat. The state (solid, liquid, or gas) does not change if the molecular structure does not. Sensible heat is created when the temperature changes while the molecular structure remains the same.

2.4.6 Latent heat

The energy (heat) necessary for a substance (state) to undergo a phase change is known as latent heat. Heat of fusion is the term used when the state is changed from solid to liquid. Heat of vaporisation is the term used to describe a transition from a liquid to a gas. When a substance receives heat after reaching the point at which it transforms into a different state, the temperature cannot rise but instead simply transforms. For instance, boiling water won't ever reach a temperature of 100 °C if you keep applying heat to it. Latent heat is typically expressed in terms of calories/gm or kilojoules per kg (kJ/kg), depending on the substance.

2.4.7 Specific Heat

The amount of heat required to raise a substance's temperature by one unit per unit of mass of that substance is known as its specific heat. The quantity fluctuates greatly based on the substance's temperature prior to applying heat. For instance, it takes one calorie to raise the temperature of water by one degree to that of room temperature, but only 0.5 calories to raise the temperature of ice by one degree to that of -5 degrees. Atmospheric pressure also affects specific heat. Lower atmospheric pressure results in a lower specific heat for the same substance. The examples that follow hold true at a pressure of one atmosphere and a temperature of 25 degrees.

2.4.8 Wet steam

Wet steam is characterized by a vapor quality ranging from zero to 1. When the vapor quality is equal to 0, it is the saturated liquid state (single-phase). On the other hand, when the vapor quality is equal to 1, it is referred to as the saturated vapor state or dry steam. A specified volume of dry steam is the amount of one kilogramme of dry steam at a specified pressure. The letter Vs is used to represent it. In steam tables, its values for various pressures are provided. With an increase in pressure, Vs loses value. In simple words, wet steam is defined as saturated steam that comprises equally dispersed saturated water particles in saturated vapour.

2.4.9 Dryness fraction

Wet steam is associated with the concept of dryness fraction. It is described as the proportion of the quantity of dry steam that is present to the mass of steam that contains it. It is typically represented by the letters " x " or "q." If m_w = Weight of water particles in suspension in the steam considered and m_s = Mass of dry steam included in the steam considered, then

$$x = \frac{m_s}{m_s + m_w}$$

Therefore, if there are 0.1 kg of water particles and 0.9 kg of dry steam in 1 kilogramme of wet steam, then $x = 0.9$.

2.4.10 Wetness fraction

It can be defined as a fraction expressing the ratio of the weight of free water particles to that of the whole in a quantity of wet steam. Also, it is the ratio of the weight of water/ moisture in suspension in a wet steam sample to the total weight of wet steam. It is calculated by subtracting x from 1.

$$i.e\ wetness\ fraction, (1 - x) = \frac{W}{W_d + W}$$

W_d = Weight of dry steam in 1 kg of wet steam,

W = Weight of water in suspension in 1 kg of wet steam

Wetness fraction is 1 for saturated water.

2.4.11 Saturated steam, superheated steam, degree of superheat

Dry saturated steam forms when wet steam is heated further and no water droplets are suspended in it. The dry saturated steam acts virtually like a perfect gas since it has absorbed all of its latent heat.

Steam is referred to as superheated steam when it is heated after being dry and saturated, and the heating procedure is referred to as superheating. Superheating is always done under a steady amount of pressure. The specific heat of superheated steam at constant pressure (Cps), whose value varies from 2.0 to 2.1 kJ/kg K depending on pressure and temperature can be used to compute the additional heat delivered to the steam during superheating.

If T_{sup}, and T_s are the wet or dry superheated steam temperatures in K, then ($T_{sup} - T_s$) is referred to as the degree of superheat.

Numerical 2.1: Calculate the enthalpy of 1 kg of steam at a pressure of 8 bar and dryness fraction of 0.8. How much heat would be required to raise 2 kg of this steam from water at 20° C?

Solution: Given: p = 8 bar and dryness fraction = 0.8

Using steam tables, the value of enthalpy of 1 kg of steam corresponding to a pressure of 8 bar is $h_f = 720.9$ kJ/kg and $h_{fg} = 2046.5$ kJ/kg

The enthalpy of 1 kg steam, $h = h_f + x\,h_{fg} = 720.9 + 0.8 \times 2046.5 = 2358.1$ kJ Ans.

The value of total heat to raise 1 kg of steam from water at 0° C was found to be 2358.1 kJ. Now the heat required to raise 2 kg of this steam from water at 20° C can be calculated by subtracting value of heat already present in the steam from total heat required to raise 1 kg of steam from water at 0° C and multiply by 2.

So, Heat already in water = 4.2 × 20 = 84 kJ

Therefore, Heat required for 1 kg of steam = 2358.1 - 84 = 2274.1 kJ

Heat required for 2 kg of steam = 2 × 2274.1 = 4548.2 kJ Ans.

Numerical 2.2: 85 kilogrammes of dry steam and 1.25 kg of water are suspended in the mixture. Find the steam's dryness fraction.

Solution: Given:

Weight of dry steam = 85 kg

Weight of water in suspension = 1.25 kg.

Now, Total weight of wet steam = 85 + 1.25 = 86.25 kg.

Now, Dryness fraction = Weight of dry steam/ Total weight of wet steam = 85/86.25 = 0.985.

Numerical 2.3: Calculate how much heat need be applied to 2 kg of water at 25°C in order to turn it into steam at 5 bar and 0.9 dry.

Solution: Given: Mass of water to be converted to steam, $m_w = 2$ kg
Temperature of water, $t_w = 25°C$
Pressure = 5 bar
Dryness fraction of steam = 0.9
Using steam tables, enthalpy at 5 Bar, $h_f = 640.1$ kJ/kg; $h_{fg} = 2107.4$ kJ/kg
Enthalpy of 1 kg of steam (above 0°C); $h = h_f + x h_{fg} = 640.1 + 0.9 \times 2107.4 = 2536.76$ kJ/kg
Sensible heat associated with 1 kg of water = $m_w \times c_{pw} \times (t_w - 0)$
$= 1 \times 4.18 \times (25-0) = 104.5$ kJ
Net quantity of heat to be supplied per kg of water = 2536.76-104.5 = 2432.26 kJ
Total amount of heat to be supplied = 2 × 2432.26 = 4864.52 kJ.

2.5 Enthalpy

A thermodynamic property known as enthalpy, or simply H, is the sum of internal energy and the pressure-volume product i.e.

$$H= U+PV$$

Enthalpy changes occur throughout heating and cooling processes, and the change in enthalpy H is significant.

$$\Delta H = \Delta U + \Delta pV$$

$$\Delta H = \Delta U+ p\Delta V+V\Delta p$$

Where Δ denotes the final value minus the starting value. When a process occurs at constant pressure, or when $\Delta p = 0$, then

$$\Delta H = \Delta U+ p\Delta V$$

As per the first law of thermodynamics,

$$q = \Delta U + p\Delta V, \text{ and}$$

$$q= \Delta H$$

The enthalpy change for a process occurring at constant pressure is equal to the heat absorbed or evolved. If the enthalpy change is positive, heat is absorbed, and the reaction is endothermic. A reaction is referred to as exothermic if it results in a negative enthalpy change, which causes heat to be released. Due to the fact that many processes occur under a constant pressure, enthalpy is frequently equated with heat content (figure 2.8).

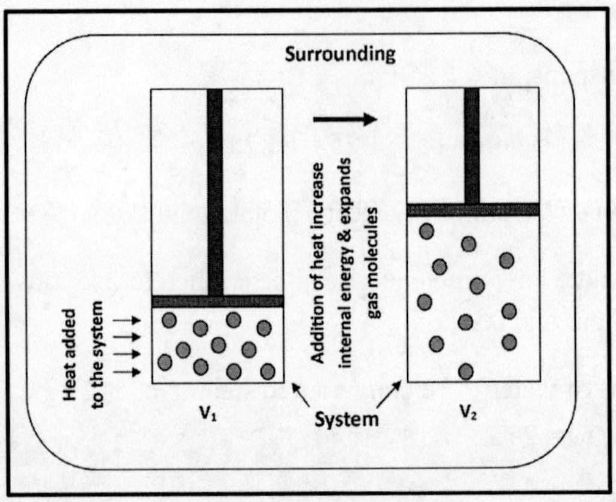

Figure 2.8 Enthalpy

Numerical 2.4: How much heat would it take to turn 4.4 kg of water at 30°C into steam at a temperature of 250°C and a pressure of 6 bar? Consider the 2.2 kJ/kg K specific heat of superheated steam.

Solution: Given: Mass of water, m = 4.4 kg, Temperature of steam, t_{sup} = 250°C

Pressure of steam, p = 6 bar, Temperature of water = 30°C

Specific heat of steam, c_{ps} = 2.2 kJ/kg

Using steam tables, the corresponding values at 6 bar and 250°C

$t_s = 158.8°C$, $h_f = 670.4$ kJ/kg, $h_{fg} = 2085$ kJ/kg

Now, Enthalpy of 1 kg superheated steam from 0°C,

$h_{sup} = h_f + h_{fg} + c_{ps}(T_{sup} - T_s) = 670.4 + 2085 + 2.2(250 - 158.8) = 2956$ kJ

Amount of heat already with 1 kg of water = 1 × 4.18 × (30-0) = 125.4 kJ

Net amount of heat required to be supplied per kg = 2956-125.4 = 2830.6 kJ

Total amount of heat required = 4.4 × 2830.6 = 12454.6 kJ.

Numerical 2.5: Calculate the mass of 0.15 m^3 of wet steam with a dryness fraction of 0.8 and a pressure of 4 bar. Determine the heat of 1 m^3 of steam as well.

Solution: Volume of wet steam, v = 0.15 m^3

Pressure of wet steam, p = 4 bar

Dryness fraction, $x = 0.8$

At 4 bar. From steam tables,

$v_g = 0.462$ m3/kg, $h_f = 604.7$ kJ/kg, $h_{fg} = 2133$ kJ/kg

∴ Density $= 1/x.v_g = 1/0.8 \times 0.462 = 2.7056$ kg/m3

Mass of 0.15 m^3 of steam = 0.15 × 2.7056 = 0.4058 kg. (Ans.)

Total heat of 1 m^3 of steam which has a mass of 2.7056 kg = 2.7056 h

(where h is the total heat of 1 kg of steam)

$= 2.7056 (h_f + x\, h_{fg}) = 2.7056 (604.7+0.8 \times 2133) = 6252.9$ kJ.

2.6 Internal energy

At a particular pressure p in kPa, the internal energy of steam (u) is the sum of the sensible heat and internal latent heat that is actually heat energy which is held in steam above the freezing point of water.

Internal energy of steam = Sensible heat + Internal latent heat

Total heat of steam = Sensible heat + Internal Latent heat + External work done

$H= U+pV$

$U= H-pV$

$U_g = H_g - pV_g$ (for dry steam)

$U_w = (H_f + x\, h_{fg}) - p(x\, V_s)$ (for wet steam)

$U_w = x\, u_{fg}$

$U_{sup} = H_g + C_{ps}(T_{sup} - T_{sat}) - pV_g \times T_{sup}/T_{sat}$ (for superheated steam)

Here H_f= sensible heat; H_g= enthalpy of dry steam; H_s= enthalpy of wet steam;

The total amount of a system's microscopic forms of energy is known as internal energy as shown in figure 2.9. It is correlated with the molecular structure and the level of molecular activity and can be thought of as the sum of the kinetic and potential energies of the molecules. Internal energy is a broad term for the assembly of a number of energy forms. For example, sensible energy is the portion of a system's internal energy that is connected to the kinetic energies of the molecules. Here, the term "kinetic energy" refers to a variety of operations, including rotations of polyatomic molecule atoms around an axis and the movement of gas molecules across space at a specific velocity. The molecules of a solid or liquid can be given enough energy to overcome these molecular forces and break out, converting the substance into a gas. The term "latent energy" refers to the internal energy connected to a system's phase change. Chemical energy is the intrinsic force connected to an atom's bond in a molecule. Thermal energy is the phrase used to describe the energy connected to the internal heat content.

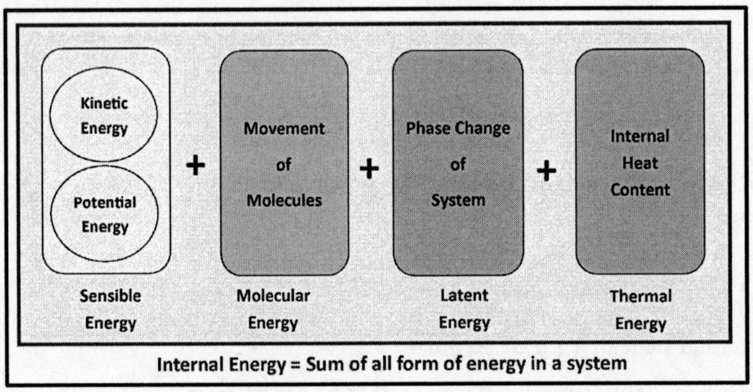

Figure 2.9 Internal Energy

2.6.1 Internal and External latent heat

A substance's volume always increases as its condition changes. As a result, we may conclude that external work is always done during a state transition. Thus, the given latent heat (L) serves two functions. The first portion is utilised to perform internal work, i.e., to work against the intermolecular force of attraction in order to increase the distance between the molecules. This is known as internal latent heat (L_i). The second portion is employed to undertake external work, i.e., to raise the volume of the gas by working against the external atmospheric pressure. This is known as external latent heat (L_e).

Here, the external latent heat represents the amount of heat that must be added to overcome the external pressure in pounds per square inch as the volume increases whereas the internal latent heat is used to overcome the resistance of the molecular forces that oppose the change in state of the molecules brought about by the formation of steam.

Numerical 2.6: A paddle wheel stirs a heated fluid that is contained in a rigid tank while cooling it. The fluid has an initial internal energy of 800 kJ. The fluid loses 500 kJ of heat throughout the cooling process, and the paddle wheel exerts 100 kJ of force on the fluid. Calculate the fluid's internal energy at the end.

Solution: Let us suppose the volume of a rigid tank is constant, and thus there is no moving boundary work. Also, heat is lost from the system, and shaft work is done on the system. Using the energy balance on the system, we get

$\Delta E = E_{in} - E_{out}$

The tank is stationary and thus the kinetic and potential energy changes are zero, $\Delta KE = \Delta PE = 0$. Therefore, $\Delta E = \Delta U$

Using this, the expression becomes $W_{in} - Q_{out} = \Delta U = U_2 - U_1$

$100 \text{ kJ} - 500 \text{ kJ} = U_2 - 800 \text{ kJ}$

$U_2 = 400 \text{ kJ}$

Numerical 2.7: At a pressure of 10 bar and a temperature of 300°C, determine the internal energy of 1 kilogramme of superheated steam. Determine the change in internal energy if this steam expands to 1.5 bar and 0.9 dryness.

Solution: Given, pressure = 10 bar and t_{sup} = 300°C

Using Steam table, internal energy, u_1 for superheated steam = 2793.2 kJ/kg

For dry saturated steam at pressure = 1.5 bar, the values are

$u_f = 466.92$ kJ/kg

$u_{fg} = 2052.7$ kJ/kg,

Now, $u_2 = u_{f2} + x_2 u_{fg2} = 466.92 + 0.9 \times 2052.7 = 2314.35$ kJ/kg

Change in internal energy, $\Delta u = u_2 - u_1 = 2314.35 - 2793.2 = -478.85$ kJ/kg

Numerical 2.8: 2 kg of steam are present in a vessel with an 8 bar pressure. To lower the quality of the steam in the vessel to 70%, determine the quantity of heat that must be rejected.

Solution: Given: mass = 2 kg; p_1 = 8 bar; x_2 = 70% = 0.7

Using steam tables, corresponding to a pressure of 8 bar are

$h_{g1} = 2767.5$ kJ/kg ; and $v_{gi} = 0.24 m^3/kg$

Now, Volume of vessel = $2 \times 0.24 = 0.48 m^3$

Initial internal energy $(u_1) = h_{g1} - 100 p_1 v_1$, $v_{gi} = 2767.5 - 100 \times 8 \times 0.24 = 2575.5$ kJ/kg

The volume occupied by the steam will now be equal to the volume of the vessel after disregarding the amount of water in the wet steam (with a dryness fraction of 0.7).

$0.48 = 2 \times 0.7 \times v_{g2} = 1.4 v_{g2}$ or $v_{g2} = 0.343 m^3/kg$

From steam tables, corresponding to a specific volume of $0.343 m^3/kg$, we find that

p_2 = 5.5 bar; h_f = 655.8 kJ/kg; and h_{fg} = 2095.9 kJ/kg

Therefore, Final internal energy per kg of steam, $u_2 = h_f + x_2 h_{fg} - 100 p_2 x_2 v_{g2}$

$= 655.8 + 0.7 \times 2095.9 - 100 \times 5.5 \times 0.7 \times 0.343$

$= 2123\text{-}132 = 1991 \text{kJ/kg}$

and heat rejected = $u_1 - u_2 = 2575.5 - 1991 = 584.5$ kJ/kg

Now the heat rejected by 2 kg of steam = $2 \times 584.5 = 1169$ kJ

2.7 Entropy

Entropy can be considered a measure of molecular disorder or unpredictability. The positions of molecules become less predictable as a system grows more disordered, and entropy increases. As a result, it is not surprising that a substance's entropy is lowest in the solid phase and largest in the gas phase. The molecules of a substance continuously oscillate about their equilibrium locations in the solid phase, but they cannot move relative to one another, and their position at any time can be predicted with high confidence. However, in the gas phase, molecules move at random, collide, and change orientation, making it extremely impossible to anticipate the microscopic state of a system at any instant (figure 2.10).

Entropy is a thermodynamic characteristic of a substance that rises with the addition of heat and drops with the removal of heat. Although entropy itself cannot be quantified, entropy change may be. The amount of heat absorbed or rejected divided by the substance's absolute temperature represents the change in entropy mathematically in a reversible process.

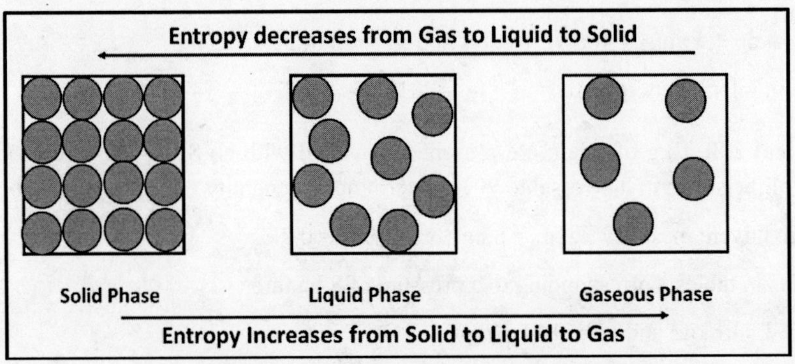

Figure 2.10 Representation of entropy change

According to second law of thermodynamics, all those processes are possible for which change in entropy for universe is greater than or equal to zero i.e. entropy of universe can never decrease and this is also known as principle of increase of entropy.

Diagram of Entropy universe

$dS \geq đQ/T$

$(dS)_{universe} \geq đQ/T$

Universe is isolated, therefore $đQ/T = 0$

So, $(dS)_{universe} \geq 0 = (dS)_{system} + (dS)_{surrounding} \geq 0$

Here dS is change in entropy and đQ/T is heat transfer.

2.7.1 Entropy of Wet Steam

It is the total of the partial evaporation entropy and the saturated water entropy. Consider 1 kg of wet steam at an absolute pressure p bar.

The entropy of wet steam can be calculated as

$s_{wet} = s_f + x s_{fg}$

When steam is dry saturated, then

Dryness fraction is 1and $s_g = s_f + s_{fg}$

Here x = dryness fraction of steam; T_s = temperature of formation of steam; s_f = the specific enthalpy of saturated water; s_{fg} = change in entropy during evaporation

2.7.2 Entropy of Dry Steam

It is the sum of entropy of saturated water and the entropy of evaporation.

Mathematically,

$s_g = s_f + s_{fg}$ (kJ/kg K)

2.7.3 Entropy of Superheated Steam

Let 1 kg of dry saturated steam to be heated to T_{sup} from T_s (steam saturation temperature in kelvin). The change of entropy during superheating at constant pressure will be equal to p if specific heat at constant pressure equals C_{ps}.

Entropy of superheating = $\int_{Tsat}^{Tsup} (\frac{\delta q}{T})$

$= C_{ps} \int_{Tsat}^{Tsup} (\frac{dT}{T}) = C_{ps} \frac{T_{sup}}{T_{sat}}$

The total entropy of superheated steam can be calculated as

S_{sup} = Entropy of dry steam + Entropy of superheating

$S_{sup} = S_f + S_{fg} + C_{ps} \ln(\frac{T_{sup}}{T_{sat}})$

Numerical 2.9: A rigid insulated tank contains 1.5 kg of air at 1 bar pressure, 300 K temperature. Through a paddle-wheel mechanism, 18 kJ of work is done on the gas during the entire process. Calculate the tank's final temperature, air pressure, and entropy change. Assume that the specific heats of air remain constant.

Solution: Given, mass of air= 1.5 kg; p1 = 1 bar = 100 kPa; T_1 = 300 K

Work done on the system (W) = –18 kJ; V = constant and Q = 0

Let C_p = 1.005 kJ/kg K and C_v = 0.716 kJ/kg K.

a) Heat transfer and boundary work are zero in an insulated, rigid tank. But through a paddle-wheel mechanism, 18 kJ of work is done on the system. So, as per first law of thermodynamics, $Q = W_{paddle} + \Delta U$

$0 = -18$ kJ + DU

$\Delta U = 18$ kJ

Also, $\Delta U = m\, C_v\,(T_2 - T_1)$; $18 = 1.5 \times 0.716 \times (T_2 - 300)$

or $T2 = 316.76$ K = 43.76°C

b) for isochoric (constant volume) process

$p_2/p_1 = T_1/T_2$

$p_2 = (316.76/300) \times 1 = 1.0588$ bar

c) Entropy change can be calculated as: $\Delta S = mC_v \ln(T_2/T_1) + mR \ln(V_2/V_1)$

As the process is isochoric, So, V= constant

Therefore, $\Delta S = mC_v \ln(T_2/T_1)$

$\Delta S = 1.5 \times 0.716 \times \ln(316.76/300)$

$\Delta S = 0.0584$ kJ/K

Numerical 2.10: At a pressure of 12 bar and a temperature of 250°C, calculate the entropy of 1 kg of superheated steam. Consider the superheated steam's specific heat to be 2.1 kJ/kg K.

Solution: Given, m = 1 kg;

p = 12 bar;

$T_{sup} = 250 + 273 = 523$ K;

$c_{ps} = 2.1$ kJ/kg K

Using steam tables; At 12 bar.,

$Ts = 188 + 273 = 461$ K,

$h_{fg} = 1984.3$ kJ/kg

Now, entropy of 1 kg of superheated steam,

$S_{sup} = c_{pw} \log_e T_s/273 + h_{fg}/T_s + c_{ps} \log_e T_{sup}/T_s$

$S_{sup} = 4.18 \log_e (461/273) + 1984.3/461 + 2.1 \log_e (523/461)$

$S_{sup} = 2.190+4.304+0.265 = 6.759$ kJ/kg

2.8 Mollier Chart

Dr. Mollier proposed a method of plotting total heat with entropy in 1904, and his diagram is more extensively used than any other entropy diagram, since work done on vapour cycles may be scaled directly as a length from this diagram, whereas it is represented by an area on the T-s diagram (figure 2.11).

Entropy or S is shown on the abscissa of the Mollier diagram, whereas enthalpy H is shown on the ordinate. It may also be known as an H-diagram. For pressures up to 1000 bar and temperatures up to 800 °C, it shows the characteristics of water and steam. It provides incremental values for the specific volume, specific entropy, specific enthalpy, and dryness percentage at various pressures and temperatures. A Mollier diagram is very useful for directly anticipating the states of steam during direct throttling and isentropic processes, as well as during compression and expansion, heating and cooling. No intricate calculations, which are necessary when utilising the steam tables, are included. Figure shown below provides an example of a Mollier chart for better comprehension.

A thick saturation line represents the 'dry and saturated state' of steam. The region below the saturation line represents steam in "wet conditions," whereas the region above the saturation line represents superheated steam. In wet and superheated environments, lines of constant temperature and constant dryness percentage are drawn. It is important to remember that the lines of constant pressure are straight in the moist zone but bent in the superheated region.

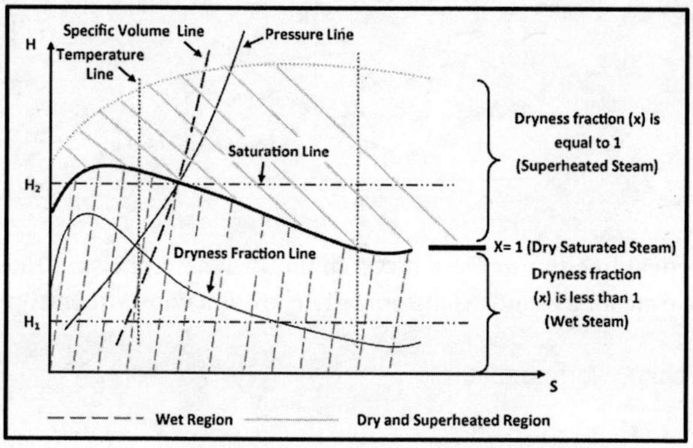

Figure 2.11 Mollier Chart

Constant pressure lines are denoted by p_1, p_2, and so on, whereas constant temperature lines are denoted by T_1, T_2, and so on. Any two independent features on the chart (e.g., p_1 and x_1 identify state 1 and h may be read off the vertical axis) are sufficient to define the state. Pressure and temperature can define the state in the superheat area (for example, p_3 and T_4 describe state 2, and h2 can be read off). The qualities at all stages throughout an isentropic transition between the two states are defined by a line of constant entropy connecting two state points 2 and 3 (figure 2.12).

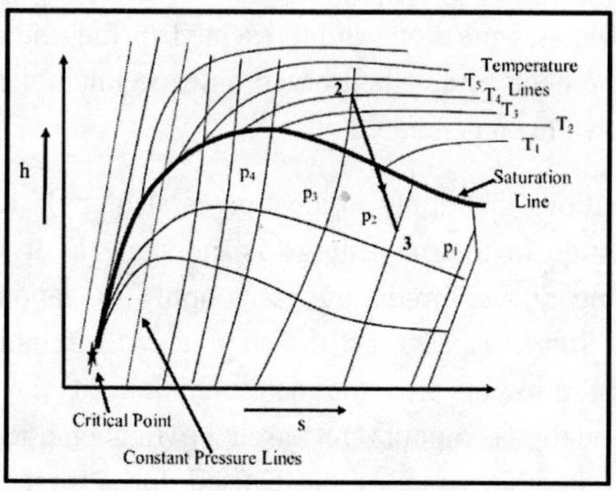

Figure 2.12 Enthalpy - Entropy Plot

Numerical 2.11: When steam is expanded isentropically from an initial pressure of 30 bar and 350° C to a pressure of 1 bar, determine the enthalpy decrease and its final condition using the Mollier chart.

Solution:
Given: $p_1 = 30$ bar; $t_1 = 350°C$; $p_2 = 1$ bar

The problem will be solved through following steps:

- ✓ First, as showed in Fig., draw a point A on the Mollier chart where the pressure line through p_1 and the temperature line through t_1 meet.

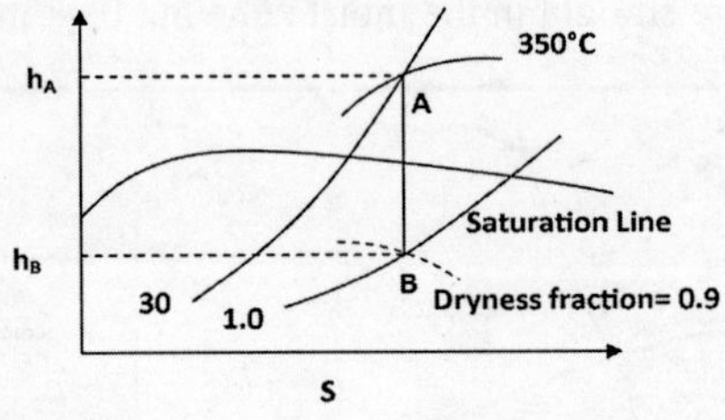

- ✓ Draw a vertical line through point A to point B, where pressure Iine p_2 (i.e., 1 bar) is met because steam expands isentropically (i.e., at a constant entropy), as illustrated in Fig.

From the Mollier chart, we find that

Enthalpy at A, $h_A = 3120$kJ/kg
Enthalpy at B, $h_B = 2450$kJ/kg
Now, the enthalpy during process (h)= $h_A - h_B$

$$h = 3120-2450 = 670 \text{ kJ/kg}$$

2.9 Various Thermodynamic Processes

A vapour is a combination of gas and suspended liquid particles that can be liquefied by minor changes in pressure or temperature. When a vapour is heated or expanded, its dryness fraction varies continually owing to evaporation of its liquid content or condensation of the gaseous component of the vapour. The pressure, volume, and temperature of vapour are not linked by a straightforward connection, as represented in the characteristic equation for an ideal gas. Equations have been established based on experimental data, and tables including vapour characteristics have been generated.

The general energy equations that apply to ideal gases also apply to vapour, and the method for determining the change in internal energy is the same as it was for gases. The same procedures for expanding and compressing gases apply to vapours as well, although the outcomes may differ. Some aspects of the vapour will remain constant during each thermodynamic process, allowing for the determination of the end state. Because the equations for work done by a vapour are based on mathematical rules of expansion or compression, they will be the same as those used for a perfect gas. Constant volume (isochoric), constant pressure (isobaric), constant temperature (isothermal), hyperbolic, polytropic, constant entropy (isentropic), and constant enthalpy (throttling) are the different thermodynamic processes for vapour. These processes are discussed in detail in subsequent sections.

2.9.1 Constant volume (isochoric) process

An isochoric process is a thermodynamic process where the closed system's volume stays constant (V=constant). It explains the non-deformable behaviour of the gas inside the container. The transfer of heat into or out of the system only affects the system's internal energy (the temperature) since the volume stays constant. The Greek terms "iso" and "choric," which both imply "space" or "volume," were combined to create the English word "isochoric."

Suppose one kg of wet steam at initial pressure p_1 and dryness fraction x_1 is heated at constant volume to final pressure p_2, then its dryness fraction will change. Let x_2 be the final dryness fraction of the steam. Since volume remains constant,

So, Volume before heating = Volume after heating

Let V_{g1} is specific volume of dry saturated steam corresponding to p_1 and V_{g2} is specific volume of dry saturated steam corresponding to p_2

Now, initial volume of wet steam can be, $V_1 = x_1 V_{g1}$

Final volume of wet steam, $V_2 = x_2 V_{g2}$

As the volume before and after heating are equal. So, $V_1 = V_2$

$x_1 V_{g1} = x_2 V_{g2}$ i.e., $x_2 = x_1 V_{g1} / V_{g2}$

The important terms related to isochoric process are:

- Work done during isochoric process

W= P.dV

As the volume is constant (so, dV = 0)

Therefore, W = 0

- Change in internal energy of steam

$du = u_2 - u_1$

If steam releases heat i.e. cooling down then du will be negative which means decrease in internal energy.

- Heat Supplied or Released

According to first law of thermodynamics,

$q = du + w \rightarrow q = (u_2 - u_1) + w$

But for isochoric process, w=0.

Therefore, $q = u_2 - u_1$ kJ/kg

The negative q means heat is released.

2.9.2 Isobaric Process

Consider 1 kg of wet steam being heated at constant pressure. As the pressure of steam remains constant its specific volume will also remain constant.

Let P = Pressure of steam in N/m^2, x_1 = Initial dryness fraction of steam, x_2 = Final dryness fraction of steam

Since the pressure (p) of steam is constant, therefore the specific volume of dry steam (Vg) corresponding to value of pressure in steam table will also remain constant, i.e. $v_{g1} = v_{g2} = v_g$.

As we know, initial volume = final volume

The important terms related to isobaric process are:

- Work done during isobaric process

W= P.dV

Here P is pressure and dV is change in volume

$W = 100p\,(v_2 - v_1)$ kJ/kg {pressure is in bar}

- Change in internal energy of steam

We know that initial internal energy of steam,

$u_1 = h_1 - 100\,pV_1$ kJ/kg

final internal energy of steam, $u_2 = h_2 - 100\,pV_2$ kJ/kg

Therefore, Change in internal energy, $du = (u_2 - u_1)$ kJ/kg

- Heat Supplied or Released

According to first law of thermodynamics,

$q = du + w \rightarrow q = (u_2 - u_1) + w$

$q = \{(h_2 - 100\,pV_2) - (h_1 - 100\,pV_1)\} + 100p\,(v_2 - v_1)$

$q = (h_2 - h_1)$ kJ/kg

The relation shows that during heat supplied or released by steam is equal to change in enthalpy.

2.9.3 Isothermal Process

For moist steam, stated in the constant pressure process, a constant temperature process is also a constant pressure process. When steam is superheated, it behaves roughly like a perfect gas and follows the hyperbolic law at constant temperature. As a result, the isothermal process will be a hyperbolic process for superheated steam, with the product of pressure and volume remaining constant. Once the steam is superheated, it behaves like a perfect gas.

In other words, the superheated steam follows the law pv/T = Constant or pv = Constant

The superheating of steam at constant temperature may be regarded as hyperbolic.

2.9.4 Hyperbolic Process

As discussed in previous process, when superheated steam acts like a perfect gas, hyperbolic process may be seen as an isothermal process in this situation. Consider 1 kg of wet steam being heated hyperbolically from initial to final state. The assumptions are same as assumed in isochoric process.

Now, initial volume of wet steam can be, $V_1 = x_1 V_{g1}$

Final volume of wet steam, $V_2 = x_2 V_{g2}$

As the steam is heated hyperbolically,

So, $p_1V_1 = p_2V_2$

$p_1(x_1V_{g1}) = p_2(x_2V_{g2})$

$x_2 = \{p_1(x_1V_{g1})\}/(p_2V_{g2})$

In case if $x_2 < 1$, then steam is wet.

if $x_2 = 1$, then steam is dry saturated.

If $x_2 > 1$, then steam is superheated.

For superheated steam, $V_{sup} = \{p_1(x_1V_{g1})\}/(p_2)$

The important terms related to hyperbolic process are:

- Change in internal energy of steam

  We know that initial internal energy of steam,

  $u1 = h_1 - 100\ p_1V_1$ kJ/kg

  Final internal energy of steam, $u_2 = h_2 - 100\ p_2V_2$ kJ/kg

  As we know in hyperbolic process, $p_1V_1 = p_2V_2$

  Therefore, Change in internal energy, $du = (u_2 - u_1)$ kJ/kg or $(h_2 - h_1)$ kJ/kg

- Heat Supplied or Released

  According to first law of thermodynamics,

  $q = du + w \rightarrow q = (u_2 - u_1) + w$

2.9.5 Isentropic Process

Isentropic or constant entropy processes are reversible adiabatic processes (i.e. frictionless adiabatic processes). However, when there is friction in the process, it is referred to be an irreversible adiabatic process. Consequently, entropy increases in an irreversible adiabatic process. In a reversible adiabatic process or isentropic process having constant entropy, no heat transfers through cylinder walls and work is done by the piston through expansion of steam.

Consider 1 kg of wet steam at initial pressure p_1 and dryness fraction x_1 is being heated isentropically to final pressure p_2, then its dryness fraction will change. Let x_2 be the final dryness fraction of the steam. Let V_{g1} is specific volume of dry saturated steam corresponding to p_1 and V_{g2} is specific volume of dry saturated steam corresponding to p_2.

The initial entropy of steam (before expansion) is given by $s_1 = s_{f1} + x_1\ s_{fg1}$

Final entropy of steam (after expansion) is given by $s_2 = s_{f2} + x_2\, s_{fg2}$

As in isentropic process, entropy before expansion = entropy after expansion

Here x = dryness fraction of steam; T_s = temperature of formation of steam; s_f = the specific enthalpy of saturated water; s_{fg} = change in entropy during evaporation

Here we can obtain the value of x_2 which will tell us about the nature of steam at final stage.

In case if $x_2 < 1$, then steam is wet.

if $x_2 = 1$, then steam is dry saturated.

If $x_2 > 1$, then steam is superheated.

The important terms related to isentropic process are:

- Work done during isentropic process

From the first law of thermodynamics,

$$q = du + w$$

As the process is isentropic,

So, q=0; therefore, 0=du + w

W= -du → w= u_1-u_2 which means work done in an isentropic process is equal to the change in internal energy.

- Change in internal energy of steam

We know that initial internal energy of steam,

$$u_1 = h_1 - 100\, p_1 V_1 \text{ kJ/kg} \qquad \{\text{here, } v_1 = x_1 v_{g1}\}$$

Final internal energy of steam, $u_2 = h_2 - 100\, p_2 V_2$ kJ/kg {here, $v_2 = x_2 v_{g2}$}

The change in internal energy is

$$du = u_2 - u_1$$

- Heat Supplied or Released

As there is no heat supplied or released, so q=0.

2.9.5 Polytropic Process (pV^n=Constant)

Consider 1 kg of wet steam at initial pressure p_1 and dryness fraction x_1 is being heated polytropically to final pressure p_2, then its dryness fraction will change. Let x_2 be the final dryness fraction of the steam. Let V_{g1} is specific volume of dry saturated steam corresponding to p_1 and V_{g2} is specific volume of dry saturated steam corresponding to p_2. Let n is the Polytropic index.

Initial volume of wet steam, $v_1 = x_1\, V_{g1}$

Final volume of wet steam, $V_2 = x_2\, V_{g2}$

As the steam is heated polytropically,

$$p_1 v_1{}^n = p_2 v_2{}^n$$

$$p_1(x_1 v_{g1})^n = p_2(x_2 v_{g2})^n$$

$$\text{So, } x_2 = \{x_1 v_{g1}/x_2 v_{g2}\} \times (p_1/p_2)^{1/n}$$

Here we can obtain the value of x_2 which will tell us about the nature of steam at final stage.

In case if $x_2 < 1$, then steam is wet.

if $x_2 = 1$, then steam is dry saturated.

If $x_2 > 1$, then steam is superheated.

The important terms related to isentropic process are:

- Work done during isentropic process

  $w = 100\ (p_1 v_1 - p_2 v_2)/n-1$

- Change in internal energy of steam

  We know that initial internal energy of steam,

  $u_1 = h_1 - 100\ p_1 V_1$ kJ/kg {here, $v_1 = x_1 v_{g1}$}

  Final internal energy of steam, $u_2 = h_2 - 100\ p_2 V_2$ kJ/kg {here, $v_2 = x_2 v_{g2}$}

- Heat Supplied or Released

  According to first law of thermodynamics,

  $q = du + w \rightarrow q = (u_2 - u_1) + w$ kJ/kg

2.9.6 Constant Enthalpy (Isenthaplic) Process

When steam is expanded via a narrow opening, such as the throat of a nozzle, the throttling process takes place. No work is done during this process, and the enthalpy stays constant. The fluid is driven out of the constricted opening or aperture during the throttling process by its pressure. Because of the narrowness of the aperture, the velocity of the fluid that exits is virtually nil due to frictional resistance between the fluid and the walls of the aperture. Friction turns the kinetic energy into heat. As a result, if the steam starts off being moist, it will begin to dry out as it expands. The dry steam will overheat if it is throttled. In an isenthalpic process having constant enthalpy, no heat transfers and no work is done by the piston through expansion of steam. Also, there is no change in internal energy.

Now, the enthalpy before expansion $h_1 = h_{f1} + x_1 h_{fg1}$

Enthalpy after expansion $h_2 = h_{f2} + x_2 h_{fg2}$ {for wet steam after expansion}

$= h_{f2} + h_{fg2} + c_{ps}\ (T_{sup} - T_s)$ {for superheated steam after expansion}

Now for isenthalpic process, $h_1 = h_2$

For wet steam after expansion

$h_{f1} + x_1 h_{fg1} = h_{f2} + x_2 h_{fg2}$

$x_2 = \{(h_{f1} + x_1 h_{fg1}) - h_{f2}\} / h_{fg2}$

For superheated steam after expansion

$h_{f1} + x_1 h_{fg1} = h_{g2} + c_{ps} (T_{sup} - T_s)$

Here we can obtain the value of x_2 which will tell us about the nature of steam at final stage.

In case if $x_2 < 1$, then steam is wet.

if $x_2 = 1$, then steam is dry saturated.

If $x_2 > 1$, then steam is superheated.

2.10 Steam Calorimeter

Superheated steam is used to power several of the steam prime movers. When the pressure and temperature of such steam are known, the enthalpy (total heat) may be easily calculated. However, saturated steam or wet steam is frequently provided. When the pressure is known, measuring its temperature simply confirms that the steam is saturated or moist. It does not provide any information about the quality or enthalpy of steam. To measure the quality or dryness fraction of the steam, some devices are used, which are known as steam calorimeters. The various types of calorimeters are discussed in subsequent sections.

2.10.1 Barrel Calorimeter

The Barrel Calorimeter is made up of a cold water-filled copper vessel. Any heat transmission from or to the system is prevented by the insulation that separates the copper vessel from its surroundings. The vessel's top is also protected by a removable insulating material. A thermometer is put into the water to record its temperature. The first step in the procedure is to open the valve in the sampling tube, allowing a steam sample from the boiler's main steam pipe to discharge into the vessel's cold water through small exit holes. Water causes steam to condense, which releases all of the latent heat and some of the sensible heat. The temperature of the water rises as a result of this heat transfer. The amount of steam condensed increases the mass of water in the calorimeter. In order to minimise experiment error, a substantial amount of steam should be introduced into the calorimeter, causing a noticeable rise in temperature. The steam supply has now shut down. The dryness fractions of steam can be expressed as follows, but barrel calorimeters are only accurate when huge amounts of steam and cold water are employed. The measured end temperature of the water may be lower due to heat conduction and radiation losses.

Now,
Let p = pressure in main steam pipe

L = enthalpy of evaporation corresponding to pressure p,

t_s = temperature of formation of steam corresponding to pressure p,

x = Dryness fraction of steam in the main pipe,

m = mass of steam blown into the vessel

M = mass of cold water in the vessel at commencement plus the mass of water equivalent of the vessel

t_1 = initial temperature of water,

t_2 = final temperature of water (mixture), and C_p = specific heat of water

Now, Heat lost by wet steam = Heat gained by cold water

$$m[x \times L + k(t_s - t_e)] = M \times C_p(t_2 - t_1)$$

$$x \times L + k(t_s - t_2) = \left[\left(\frac{M}{m}\right) C_p(t_2 - t_1)\right]$$

Dryness fraction, $x = \left[\frac{M}{m} k(t_2 - U) - k(t_s - t_2)\right] / L$

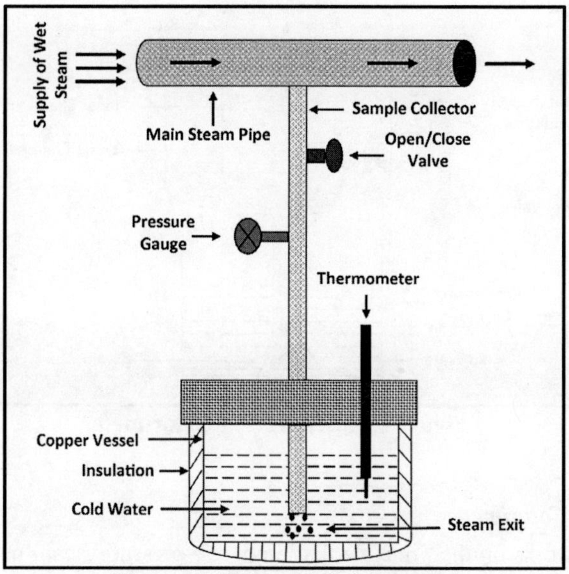

Figure 2.13 Barrel Calorimeter

2.10.2 Separating Calorimeter

The dryness fraction from separating calorimeter is calculated by separating the moisture content from the wet steam. The water separated from the moist steam is collected in an inner chamber with cross-sectional area "A." To measure the amount of water separated after and before sampling, a gauge glass with a graduated scale is installed in the inner chamber of the separating calorimeter. It also has a steam condensing chamber where dry and saturated steam can be condensed and measured. A sample tube controlled by a valve allows moist steam to enter the inner chamber of the separating calorimeter. The valve should be completely open to prevent any throttling or wiredrawing while checking for dryness. Separators or baffle plates are installed in the inner chamber of a separating calorimeter in order to separate the moisture from the steam. As a result, as steam enters the separating calorimeter, it experiences an abrupt change in direction of travel when it reaches the baffle plates. As a result, the water particles with the greatest inertia separate from the moist steam. Water flows downhill and collects at the bottom of the inner chamber of the separating calorimeter. When dry steam reaches the annular area between the separating calorimeter's inner and outer surfaces, it travels upward and descends. The separating calorimeter discharges dry and saturated steam, which is then condensed in the steam condensing chamber. The weight of the condensed steam in the steam condensing chamber is then determined.

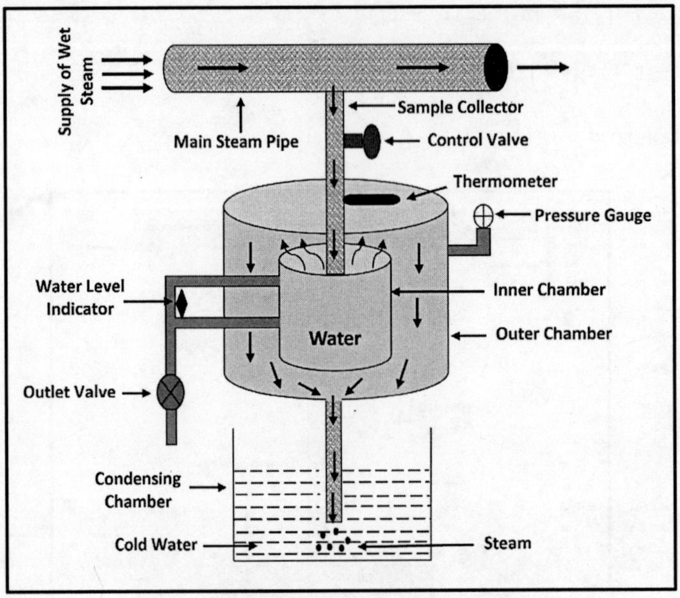

Figure 2.14 Separating Calorimeter

2.10.3 Throttling Calorimeter

This calorimeter works on the principle that moderate-pressure steam has more total heat or enthalpy than low-pressure steam. When steam passes through a small orifice, there is a drop in its pressure. As no external work is done by the steam when passing through the orifice in the throttling valve, the enthalpy of steam before throttling must be equal to the enthalpy of steam after throttling. Hence, if high-quality steam at moderately high pressure is throttled, it tends to become superheated after throttling, i.e., if the steam is fairly dry initially, the pressure drop during throttling will release sufficient heat to superheat the steam. The throttling calorimeter (fig. 2.15) consists of a chamber, which is shaped to provide a thermometer pocket in the centre. The chamber has a throttling valve (orifice) at the top through which steam enters the calorimeter. The steam expands to a pressure a little above atmospheric pressure after passing through the orifice. The steam leaves the calorimeter and passes into the atmosphere through the exhaust pipe. The sampling tube placed in the main steam pipe is connected to the calorimeter chamber. The pressure of steam before passing through the throttle valve is measured by a steam pressure gauge (not shown) attached to the main steam pipe. The pressure of steam inside the calorimeter chamber (after throttling) is measured by means of a manometer gauge, which is a glass U-tube containing mercury. The temperature of throttled steam is read by the thermometer placed in the pocket containing oil. The throttling calorimeter is designed to throttle very nearly dry saturated steam, thus, producing superheated steam whose temperature and pressure can be measured. By reading the initial and final conditions (temperature and pressure) of the steam, its dryness fraction can be measured. Throttling calorimeters will only be useful, if the steam after throttling is superheated.

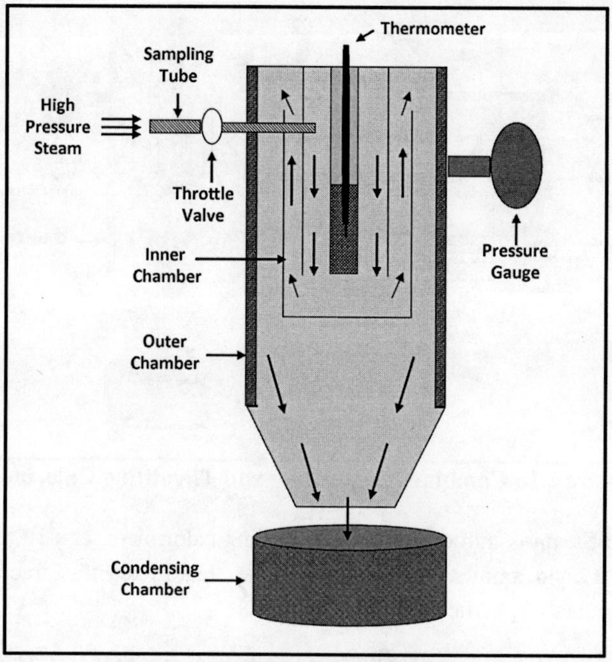

Figure 2.15 Throttling Calorimeter

2.10.4 Combined Separating and Throttling Calorimeter

The throttling calorimeter fails if the steam is not superheated after throttling, and the separating calorimeter does not produce an accurate result. It has been determined that the best method for precisely measuring steam dryness is a combination of separating and throttling calorimeters.

In this instrument, the sample of wet steam is first passed through the separating calorimeter, where most of the moisture content (water) of the steam is separated and collected. The drier sample of steam then passes to the throttling calorimeter, where it is superheated after throttling. In estimating the dryness fraction of steam supplied to the combined separating and throttling calorimeter, first the dryness fraction of the steam entering the throttling calorimeter is determined by using the following expression:

x = Mass of dry steam present in the mixture/Total mass of the mixture

$$x = \left(\frac{M}{M+m}\right) \times x_2$$

$$x = x_1 \times x_2$$

Here, $M + m$ is total mass of supplied steam and $M \times x_2$ is mass of dry steam; x_2 is dryness fraction of steam entering the throttling calorimeter

M is mass of steam passed through the throttling calorimeter after separating some water in the separating calorimeter

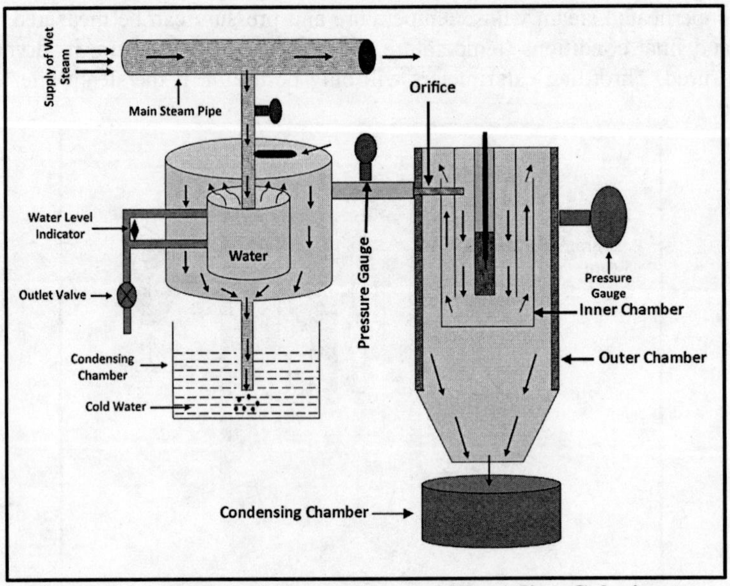

Figure 2.16 Combined Separating and Throttling Calorimeter

Numerical 2.12: Steam is introduced into a throttling calorimeter at a 10 bar pressure and is throttled to 110°C and ambient pressure. Find the steam's dryness fraction. Assume that superheated steam has a specific heat of 2.2 kJ/kg K.

Solution: Given, $p_1 = 10$ bar; after throttling $p_2 = 1.01325$ bar, $T_2 = 110$ °C; $C_{ps} = 2.2$ kJ/kg.K

Using steam table, at p_1, $h_{f1} = 762.81$ kJ/kg

$h_{fg1} = 2014.3$ kJ/kg

At p_2, $h_{g2} = 2676$ kJ/kg

$T_{sat2} = 100$°C

As the enthalpy is constant in throttling process, therefore $h_1 = h_2$

$$h_{f1} + x\, h_{fg1} = h_{g2} + C_{ps}(T_2 - T_{sat2})$$

$$762.81 + x \times 2014.3 = 2676 + 2.2\,(110 - 100)$$

$$x = (2676 + 22 - 762\ 81)/2014.3$$

Dryness fraction, $x = 0.961$

Numerical 2.13: A sample of moist steam is sent through a separating calorimeter as part of an experiment. At some point, the separator's mass of water collected was 0.2 kg, while the majority of condensed steam was discovered to be 2.5 kg. Find out what percentage of the steam that enters the calorimeter is dry.

Solution: Mass of water, $m_w = 0.2$ kg; Mass of steam, $m_g = 2.5$ kg
The dryness fraction of wet steam is calculated as

$$x = m_g/(m_g + m_w)$$

Using numerical values, we get x =25/(25 +0.2)

$$x = 0.926$$

Numerical 2.14: The dryness fraction of steam passing through a steam main at a pressure of 9 bar was measured using a combination separating and throttling calorimeter. After throttling, steam had a pressure of 1.25 bar and a temperature of 115°C. After throttling, 2.2 kg of steam condensed, and 0.20 kg of water was collected in the separator. Calculate the primary steam's dryness fraction. Consider the 2.1 kJ/kg K specific heat of superheated steam.

Solution: Given: $m_w = 0.2$ kg; $m_g = 2.2$ kg; $p_1 = 9$ bar; $p_2 = 1.25$ bar; $T_2 = 115$ °C; $C_{ps} = 2.1$ kJ/kg
The dryness fraction x1 of wet steam from separating calorimeter $x_1 = m_g/(m_w + m_g)$

$$x_1 = (22/22{+}0.2) = 0\ 9167$$

Before throttling at $p_1 = 9$ bar; $h_{f1} = 742.83$ kJ/kg; $h_{fg1} = 2031.1$ kJ/kg
After throttling at $p_2 = 1.25$ bar; $T_{sat2} = 106$ °C; $h_{g2} = 2685.4$ kJ/kg
Enthalpy before throttling, $h_1 = h_{f1} + x_2\, h_{fg1} = 742.83 + x_2 \times 2031.1$
Enthalpy after throttling, $h_2 = h_{g2} + C_{ps}\,(T_2 - T_{sat2})$

$$h_2 = 2685.4 + 2.1 \times (115 - 106) = 2704.3 \text{ kJ/kg}$$

As, $h_1 = h_2$;
Therefore,

$$742.83 + x_2 \times 2031.1 = 2704.3$$

$$x_2 = (2704.3 - 742.83)/2031.1 = 0.9657$$

UNIT SUMMARY

Based on the contents covered in this unit you should remember that:

- Since steam is a vapor, it is not subject to ideal gas laws until it reaches an extremely dry state.
- The dryness fraction is the ratio of the mass of actual dry steam to the mass of steam containing it.
- The steam is divided into two major groups: dry steam and wet steam.
- The term "degree of superheat" refers to the overall increase in temperature of superheated steam above the saturation temperature.
- Under constant pressure, the saturation temperature, latent heat, and other properties of steam remain constant; however, when pressure varies, they undergo changes.
- When volume compresses at a constant temperature and pressure, saturated vapor condenses into a liquid quickly enough to maintain a constant pressure.
- The intersection of the saturated liquid and saturated vapor lines is the critical point.
- At constant temperature and pressure, saturated vapor condenses to liquid at a rate sufficient to maintain a constant pressure.
- Latent heat refers to the energy required for a material to undergo a phase change, such as from solid to liquid or liquid to gas.
- Steam calorimeter helps in determining the quality of steam.
- Understanding the steam quality in industrial settings is essential for ensuring the best possible efficiency and functionality of steam turbines and heat exchangers.

Exercises

Multiple Choice Questions

1. **What is the primary process by which steam is formed?**

   A) Condensation

   B) Evaporation

   C) Freezing

   D) Sublimation

2. **At what temperature does water typically boil and turn into steam at standard atmospheric pressure?**

   A) 50°C

   B) 75°C

   C) 100°C

   D) 125°C

3. **Which of the following conditions will increase the rate of steam formation?**

   A) Decreasing the temperature of water

   B) Increasing the temperature of water

   C) Increasing the atmospheric pressure

   D) Reducing the surface area of the water

4. **What is the term for the energy required to convert water into steam without changing its temperature?**

   A) Latent heat of fusion

   B) Sensible heat

   C) Specific heat capacity

   D) Latent heat of vaporization

5. **What does the Mollier chart primarily represent?**

   A) The relationship between pressure and temperature of gases

   B) The enthalpy and entropy of steam

   C) The specific heat capacity of liquids

   D) The relationship between viscosity and temperature

6. **On a Mollier chart, which axis typically represents enthalpy?**

   A) Horizontal axis

   B) Vertical axis

   C) Diagonal axis

   D) No specific axis; it's represented by contours

7. **Which line on a Mollier chart represents the saturation curve?**

A) The line of constant pressure

B) The line of constant temperature

C) The boundary between the liquid and vapor phases

D) The line of constant entropy

8. The quality of steam, as measured by a steam calorimeter, refers to:

A) The temperature of the steam

B) The specific enthalpy of the steam

C) The pressure of steam

D) The ratio of the mass of vapor to the mass of steam and water

9. In a steam calorimeter, the term "quality" of steam is expressed as:

A) A percentage

B) A temperature difference

C) A pressure difference

D) A volume ratio

10. Which of the following components is typically found in a steam calorimeter?

A) A pressure gauge

B) A temperature sensor

C) A liquid-water separator

D) A heat exchanger

11. In a steam calorimeter, the process of separating steam into saturated vapor and liquid is crucial for:

A) Determining the steam quality accurately

B) Calculating the temperature of the steam

C) Measuring the pressure of the steam

D) Measuring the enthalpy of the steam

12. How is the quality of steam calculated using a steam calorimeter?

A) By measuring the temperature difference between the steam and the condensate

B) By measuring the specific enthalpy of the steam

C) By measuring the ratio of the mass of condensed water to the total mass of steam

D) By measuring the total volume of steam and condensate

Answer Key:

1 B	2 C	3 B	4 D	5 B	6 B
7 C	8 D	9 A	10 C	11 A	12 C

Short and Long Answer Type Questions

❖ What does saturation pressure and saturation temperature mean?

- ❖ Describe the processes involved in generating wet steam, dry saturated steam, and superheated steam.
- ❖ What does the term "internal energy of steam" mean?
- ❖ What does "constant dryness fraction lines" mean? How is the T-S diagram drawn with these?
- ❖ Write a brief note on a generation of the Mollier chart.
- ❖ Describe the process of steam generation and provide a graphical depiction of it.
- ❖ Describe the steam throttling procedure.
- ❖ What are the disadvantages associated with the barrel calorimeter?
- ❖ Which calorimeter is used for very moist steam?
- ❖ Write the steam's internal energy expressions for both its wet and superheated states.
- ❖ Generate steam entropy expression for both wet and superheated steam conditions.

Know More

- ✓ Steam is an important part in the thermodynamic cycle. It is used by the Rankine cycle for the power generation and steam turbine uses to convert thermal energy into mechanical work.
- ✓ The phase diagram of water gives correlation between pressure and temperature and generates boundary lines between all three phases of water i.e. solid, liquid and gas. This representation shows critical point which at liquid and vapor phases are difficult to recognized.
- ✓ When steam, at specific pressure, is heated beyond its boiling point then it is turned into superheated state. This phenomenon improves the heat content of the substance and enables it to expand which in result becomes more useful for power turbines.
- ✓ The ratio of steam to the total amount of steam and water gives steam quality. It is expressed in percentage. The steam having lesser amount of water is more efficient in energy transfer.
- ✓ Steam tables offer broad thermodynamic information, including temperature, pressure, enthalpy, and entropy, for variable steam conditions. The Mollier diagram, often known as the h-s diagram, is a graphic tool that demonstrates the association between enthalpy and entropy. It is broadly used in thermodynamics for the study of steam cycles and processes.

References and Suggested Readings

- ➢ Thermal Engineering by B. R. Sarkar (McGraw-Hill Education (India) Pvt Limited).
- ➢ Thermal Engineering by R.K. Rajput (Laxmi Publications (P) Ltd).
- ➢ A Textbook of Thermal Engineering by R.S. Khurmi & J.K. Gupta (S. Chand & Company Ltd).
- ➢ Thermal Engineering by Mahesh. M. Rathore (McGraw-Hill Education (India) Pvt Limited).

Dynamic QR Code for Further Reading

1. Lectures on Steam formation and Properties of steam in Basic Mechanical Engineering by Engineering Funda.

2. Lectures of the NPTEL Course on "Property of pure substance and Steam Table" by Dr P.K. Das, Department of Mechanical Engineering, Indian Institute of Technology Kharagpur, India.

3. Lectures on "Throttling calorimeter" by Prof. Mayur. S. Modi's Modi Mechanical Engineering Tutorials.

4. Lectures on "Seperating calorimeter" by Mechanical Engineering Management YouTube channel.

3 Steam Generators

UNIT SPECIFICS

This unit presents information related to the following topics:

- Function and classification of steam boilers by using schematic diagrams.
- Comparison of water tube and fire tube boilers and discussion on Lamont and Benson boilers.
- Description of boiler mountings and accessories.
- Calculation of boiler efficiency, boiler horsepower, actual evaporation, equivalent evaporation.
- Detailed explanation of Draught systems (Natural, forced & induced) using diagrams.

This unit provides readers with a necessary understanding of steam generators or boilers. The author provides specific information on different types of boilers according to their classification. The comparative analysis of the water tube and fire tube boilers will shed light on key differences between the two. The unit includes different schematic diagrams so that readers can easily distinguish between different types of boilers. The boiler mountings and accessories are discussed in detail. The unit will help readers in calculating evaporation, equivalent evaporation, factor of evaporation, boiler horse power, and boiler efficiency using formulas. The last part of the unit includes detailed explanations of the draft system using appropriate diagrams.

A brief overview of the primary concepts is provided at the end of the unit. There are also an adequate number of multiple-choice and descriptive questions in this course. For students to use as practice, the unit contains a list of books and other resources. It is worth mentioning that certain sections feature QR codes to provide further information on different topics of interest. Readers may receive relevant and confirming information by scanning these codes. In order to give additional information on a variety of topics of interest, the unit promotes the use of QR codes and video resources. Mobile users may browse or scan QR codes to access and see these content.

RATIONALE

Steam boilers play a crucial role in several industries, serving as the core of energy generation and industrial operations. Power plants convert water into high-pressure steam to power turbines and generate electricity, making them essential for large-scale energy production. Steam boilers play a vital role in maintaining pleasant and efficient heating in residential, commercial, and industrial settings. They have a crucial function in industrial operations, such as manufacturing, by providing steam for functions like sterilizing, drying, and powering machinery.

The potential of steam boilers or steam generators makes them significant subjects for study. This unit will help the readers learn about the various types of boilers and how they work. As the unit progresses, the readers will learn about the various important terms, which will help to advance the subject. The discussed formulae and diagrams help them gain knowledge about boiler efficiency and performance parameters.

PRE-REQUISITES

Before reading this unit, the students are advised to revisit the following:
Basic Mechanical Engineering (MEPC102)

UNIT OUTCOMES

After studying this unit students will be able to:

U3-O1: Understand the boilers and their classification and function.

U3-O2: Explain Cochran, Babcock and Wilcox boilers with diagrams.

U3-O3: Comparison of water tube and fire tube boilers and description of high pressure boilers.

U3-O4: Learn about the boiler mountings, accessories and performance measuring terms.

U3-O5: Understand the types of draught system.

Unit 3 Outcomes	**EXPECTED MAPPING WITH COURSE OUTCOMES** (1- Weak Correlation; 2- Medium Correlation; 3- Strong Correlation)				
	CO-1	**CO-2**	**CO-3**	**CO-4**	**CO-5**
U3-O1	1	2	3	-	-
U3-O2	1	1	3	-	-
U3-O3	-	1	3	-	-
U3-O4	1	-	3	-	-
U3-O5	-	-	3	-	-

3.0 Steam Boilers

A boiler is a device that uses fuel combustion to produce steam from water in a closed system. It is not always necessary to undergo the process of converting water into steam on every occasion. The boilers have the potential to be utilised for heating applications as well. Generally, a boiler in a steam power plant consists of three parts: (a) a furnace in which fuel is burned; (b) surfaces for transmitting heat from the combustion products to the water; and (c) a compartment where steam may build and gather.

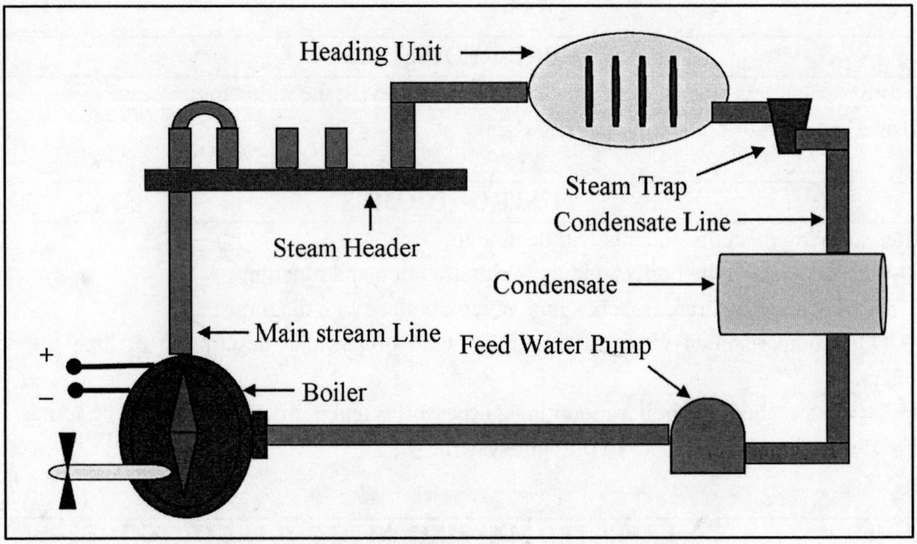

Fig. 3.1 Schematic Diagram of Steam Boiler

3.1 Function and use of steam boilers

A sealed container, commonly referred to as a boiler, is used for the heating of fluids, such as water or other chemicals, with the objective of producing steam or hot water. Boilers function by utilizing the fundamental principle of heat transfer, whereby heat produced via the process of combustion is transferred to a fluid medium in order to elevate its temperature. A range of substances, including coal, oil, natural gas, and biomass, can be used as fuel sources during the process of combustion. Subsequently, the heated fluid facilitates the passage of thermal energy to a heat exchanger, which in turn transmits the heat to a fluid, such as air or water, with the capacity to effectively transport thermal energy.

The purpose of a boiler is contingent upon its specific classification and intended use. The primary function of a hot water boiler is to provide heat to water intended for home or commercial use, whereas a steam boiler is specifically built to produce steam for applications in power plants or other industrial processes.

Moreover, steam or hot water may be utilized for many industrial applications, such as energy generation, food processing, medical equipment sterilization, garment drying, and other related uses. Numerous heating and power production systems need the use of boilers in order to function effectively.

3.2 Classification of steam boilers

The identification of a steam boiler is determined by several factors, including its structural characteristics, mobility, tube configuration, fuel type, and pressure generation. While early boilers showed comparable efficiency with modern boilers, their operation posed significant hazards due to the absence of control mechanisms. The size, fuel, and dimensions of a boiler vary based on the work it is designed to perform and the industry it serves. Electricity, wood, natural gas, coal, and fossil fuels are categorized as various types of fuel. The cost, environmental sustainability, and efficiency of fuels show regional variations. The boilers are classified into seven categories as shown in figure 3.2.

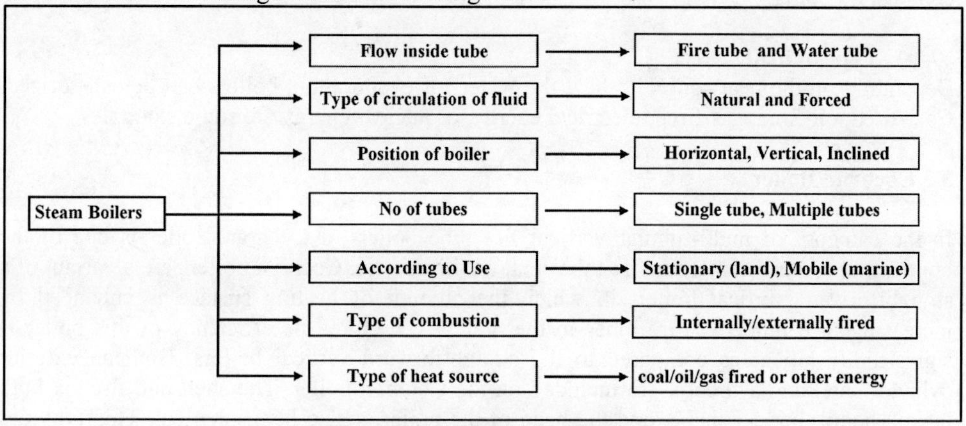

Fig. 3.2 Classification of Boilers

The boilers are classified on the following basis:

a) Flow inside tubes

On the basis of flow inside tubes, there are two types of boilers i.e. fire tube and water tube boiler. Water or steam is forced through the tubes of a water tube boiler, which heats the area outside the boiler. This category includes Babcock and Wilcox and Stirling boilers.

In a fire tube boiler, combustion products or hot flue gases flow inside the tubes, while water flows outside the tubes. The boilers like Cornish boiler, Lancashire boiler and Cochran or locomotive boiler are various examples of fire tube boilers.

b) Type of circulation of fluid

There are generally two types of flows named as natural and forced. Water is naturally circulated by natural convection currents created by the application of heat, as in the case of Lancashire and Babcock & Wilcox boilers.

When a system uses forced circulation, the fluid is either forced "once through" or managed with partial recirculation, as in the cases of LaMont, Velox, and Benson boilers.

c) Position of boiler

The position of boilers here refers to the orientation of boiler with reference to the surface. The orientation of the boiler's main axis determines whether it is horizontal, vertical, or inclined.

d) No. of tube(s)

There may be several fire/water tubes in a boiler, or there may be just one tube.

e) According to use

Both stationary (land) and mobile (marine and locomotive) boilers are used in boiler systems. Steam is produced at power plants using stationary boilers. Boilers those are easily transportable are termed as mobile.

f) Type of combustion

When combustion occurs out from the area where the water is boiling, a boiler is referred to as an external combustion boiler. When the furnace zone is entirely covered by a water-cooled surface, as in the case of the Lancashire boiler, the boiler is referred to as an internal combustion boiler.

g) Type of heat source

Depending on the heat source to heat the water for evaporation, boilers can be categorized. Coal-fired, oil-fired, gas-fired, electrical energy, or nuclear energy are some examples.

3.3 Cochran Boiler

In the category of multi-tubular vertical fire tube boilers, a Cochran Boiler is one of the boilers that contain numerous fire tubes that are horizontal. Cochran boilers are a variant of a straightforward vertical boiler, in which the amount of heating surface is enhanced by increasing the number of fire tubes in the boiler. Therefore, the efficiency of this boiler is significantly higher as compared to the straightforward vertical boilers. Both an exterior cylinder shell and a firebox are included on the Cochran boiler. The shell and firebox both have a hemispherical shape, and the head of the boiler is also hemispherical. This provides the boiler with the maximum amount of space and strength it requires to withstand the pressure that is built up inside the boiler. In addition to this, the firebox has a hemispherical crown, which helps it better withstand tremendous heat.

There are more than twenty parts and components involved in a single Cochran boiler. We decided to focus only on the most important of these parts and give a brief description of what they do and what role they carry in the boiler. Below we listed nine of them.

1. **Boiler Shell**

The boiler shell consists of steel plates moulded into a cylindrical shape and bolted or welded together. The endplates secure the shell's ends. The boiler's outer shell has to be large enough to hold both water and steam.

2. **Combustion Chamber**

Under the boiler shell, the combustion chamber burns the fuel to produce steam from the water in the shell.

3. **Grate**

The combustion chamber's grate is a platform where fuel undergoes combustion. In most cases, the grate is made of cast iron bars with gaps between them to allow air to travel through. The grate surface refers to the region of the grate where fire burns.

4. **Furnace**

The fuel is consumed in the furnace, which is located below the boiler shell and above the grate

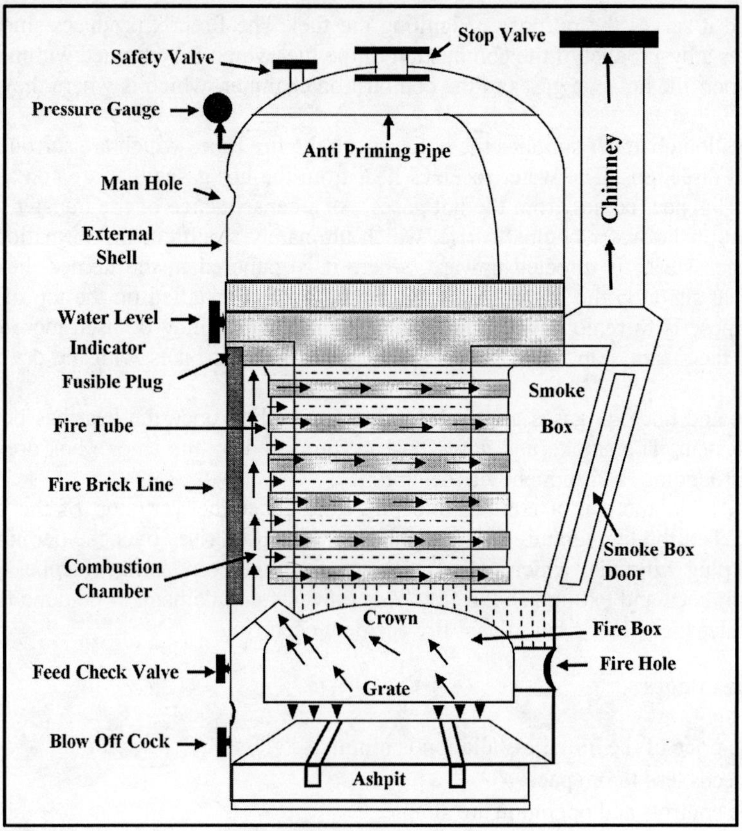

Fig. 3.3 Cochran Boiler

5. **Fire Tubes**

The horizontal tube separates the combustion chamber and the fire tubes. Several fire tubes are used to transport the combustion chamber's exhaust gases to the smoke box. These fire pipes transmit heat from hot flue gases to water.

6. **Fire Hole**

It is at the bottom of the combustion chamber for firing fuel inside the furnace.

7. **Chimney**

The smoke box and chimney are linked and are both located at the top of the boiler. These exhaust gases emerge from the fire tubes, travel through the smoke box, and exit through the chimney.

8. **Man Hole**

The manhole is for cleaning, repairing, and inspecting the boiler shell.

9. **Flue Pipe**

The short pipes that link the firebox and combustion chamber are known as flue pipes. Through the flue pipe, hot flue gases from the grate enter the combustion chamber.

➢ **Working of Cochran Boiler**

Before a Cochran boiler can begin to function, the fuel must first be placed on the grate and then brought into the firebox. A fire hole may be found on the right side of the boiler's bottom, and it serves the purpose of igniting the fuel. The firebox produces smoke and hot flue gases as a by-product of the combustion of the fuel which is contained within it. The flue pipes transport the hot flue gases to the combustion chamber, which is where they are heated up.

The combustion chamber supplies the hot gases to the fire tubes which are surrounded by the water in all directions. The water receives heat from the hot gases that are contained within the tubes. This heat comes from the hot gases. As a consequence of the transfer of heat, the temperature of the water begins to rise, which ultimately results in the formation of steam. The produced steam is directed upward, where it is gathered in the domed, hemispherical structure that sits atop the boiler. An anti-priming pipe is installed on the top of the boiler, and its purpose is to remove water from the steam so that it may be used more effectively. After that, the steam stop valve is utilised to provide the turbines with the dry steam they require.

The smoke and hot flue gases are sent into the smoke box once the heat has been rejected from the system. The smoke and gases that are produced by the smoke box are discharged into the surrounding environment via the chimney.

In addition to it, there is a combustible plug located at the very top of the combustion chamber. When the temperature in the combustion chamber rises over the permissible limit, the fusible plug will melt, which will allow water from the combustion chamber to flow into the boiler furnace and extinguish the fire. This prevents any damage to be done to the boiler and eliminates the risk of a significant fire accident occurring.

➢ Advantages

- ✓ The price of the initial installation is minimal.
- ✓ It needs less floor space.
- ✓ The controls and operation are simple.
- ✓ It is easily transportable.
- ✓ It is compatible with all gasoline types.

➢ Disadvantages

- ✓ It has a low rate of steam production.
- ✓ Limited capacity for pressure handling.
- ✓ It is challenging to maintain and evaluate.

➢ Applications of Cochran Boilers

Power stations that produce enormous amounts of steam from 500 kg/s at high pressures of about 160 bar and high temperatures of up to 550°C utilize Cochran boilers. We have mentioned some other uses for Cochran Boilers below:

- It has applications in refining units.
- It is utilised in paper production factories.
- Cochran boilers are used in the chemical processing industries.
- In addition, it is employed in several process applications across a number of sectors.

3.4 Babcock and Wilcox Boilers

It is an externally burned, multi-tubular, stationary, high pressure water tube boiler with a horizontal drum axis that runs on solid fuel and has natural circulation and draught. The boiler, sometimes referred to as the Babcock and Wilcox boiler, was developed in 1967 by George Herman Babcock and Stephen Wilcox.

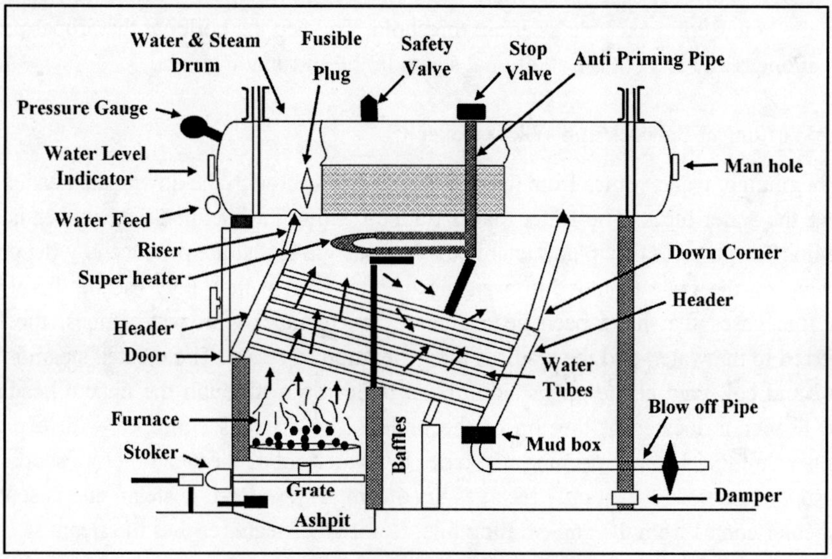

Fig. 3.4 Babcock and Wilcox Boiler

The following are the major components of a Babcock & Wilcox boiler:

- Drum: A horizontal axis drum that holds both water and steam.
- Down Take Header: This component is located at the back end of the boiler and links the water tubes to the back end of the drum. It takes water from the drum.
- Up Take Header: This component is located at the front end of the boiler and is linked to the front end of the drum. It transports steam from the water tubes to the drum.
- Water Tubes: These are the tubes through which water travels and turns into steam. It transfers heat from the hot flue gases to water. It is tilted at a 10-15° inclination with respect to the horizontal. Because of the incline, the water tubes are not entirely filled with water, and water and steam readily separate.
- Baffle Plates: Baffle plates are located between water tubes and allow hot flue gases from the furnace to move in a zigzag pattern to enhance rate of heat transfer.
- Fire Door: This is the door that is used to ignite the solid fuel in the furnace.
- Grate: It is the foundation upon which solid fuel is burned.
- Mud Collector: This device is used to collect the mud that is present in the water and is located at the bottom of the down take header.
- Feed Check Valve: This valve is used to fill the drum with water.
- Damper: This device controls the flow of air in the boiler.
- This type of boiler uses a variety of mounting systems and attachments, including:

- Super heater: Before releasing the steam from the steam stop valve, the super heater raises the saturated steam's temperature to the desired level.
- Pressure Gauge: The pressure gauge measures the steam pressure within the boiler drum.
- Water Level Indicator: This displays the amount of water in the drum.
- Safety Valve: A safety valve is a valve that opens when the pressure of steam within the boiler drum exceeds a certain threshold. To keep the boiler at the proper pressure, it opens and lets out the additional steam in the surrounding area.

➢ **Working of Babcock and Wilcox Boiler**

In the beginning, water comes from the drum and passes through the down take header before entering the water tubes. The water that is trapped within the inclined water tubes is heated by the hot flue gases. Baffle plates enable the hot flue gases that are produced by the coal that is burning on the grate to move in a zigzag manner so that they may escape the chimney. When flue gases that have been heated come into contact with water tubes, the heat is transferred to the water, and the result is the generation of steam. The created steam is forced upward and collected at the upper side of the boiler drum through the uptake header. The uptake header is located at the top of the boiler. The drum is equipped with a pipe that prevents priming. This anti-priming line takes the water out of the steam before sending it on to the super heater, so that it only receives dry steam. The water-free steam that is sent to the super heater comes from the anti-priming line. Before the steam reaches the steam stop valve, the temperature of the steam is increased to the desired amount by this component. Either the superheated steam from the steam stop valve is stored in a steam drum or it is made to strike on the steam turbine in the steam power plant so that electricity may be produced.

3.5 Comparison of water tube and fire tube boilers

The major difference between a fire tube boiler and a water tube boiler is that in a fire tube boiler, hot gases flow within the tube and water surrounds them, whereas in a water tube boiler, water flows inside the tube and hot gases surround them. The following is a comparison of the two boilers:

FIRE TUBE BOILER	**WATER TUBE BOILER**
Hot flue gases circulate within the tubes, while water flows outside.	Water circulates within the tubes, while hot flue gas flows outside.
These boilers are often fired from inside.	These boilers are fired from outside.
The maximum boiler pressure is 20 bar.	The maximum boiler pressure is 100 bar.
The fire-tube boiler produces less steam at a slower rate.	The water tube boiler produces steam at a higher rate.
Insufficient for larger power plants.	Sufficient for larger power plants.

FIRE TUBE BOILER	WATER TUBE BOILER
Because of the low pressure, there is less chance of an explosion.	Because of the high pressure, there is more chances of an explosion.
It takes up a lot of floor space for a given amount of power.	It takes up a less floor space for a given amount of power.
Construction is difficult.	Construction is simple.
Transportation is challenging.	Transportation is simple.
They require less skill to operate than a water tube boiler.	They require high skill to operate than a water tube boiler
They are tough to repair and clean due to the fact that they are internally fired.	They are easy to repair and clean as they are externally fired.
Because the fire tube is located inside the shell, they require a high shell diameter.	They require a small shell diameter.
The fire tube boiler has a lower efficiency than the water tube boiler.	The efficiency of the water tube boiler is more.
This boiler's servicing is expensive. It must be inspected on a regular basis.	They are easy to maintain as they are externally fired.
Example: Cornish Boiler, Lancashire Boiler	Example: Babcock and Wilcox Boiler

3.6 High Pressure Boilers

A high-pressure boiler is a specific type of boiler that generates power in thermal power plants by operating at pressures more than 80 bars. Steam generated by a high-pressure boiler typically ranges from1 bar to 55 bar. The main difference between low-pressure and high-pressure boilers is that low-pressure boilers run at pressures less than 1.03 bar, and high-pressure boilers work at pressures more than 1.03 bar.

3.6.1 Lamont Boiler

Thomas C. Lamont developed the Lamont Boiler, a form of forced-circulation water-tube boiler, in the early 20th century. The high-pressure steam produced by this type of boiler may be used to power turbines, motors, pumps, and other mechanical devices in a range of industrial applications. The Lamont Boiler is distinguished from other boilers by its forced-circulation design, which employs a network of pumps and pipelines to circulate water within the boiler. When compared to conventional boilers, this delivers more efficiency and enables for the production of higher-pressure steam. Moreover, the Lamont Boiler is very efficient and can generate steam from a variety of fuel sources, including coal, oil, and natural gas.

❖ Construction of Lamont Boiler

A Lamont Boiler consists of several parts which arc as follows:

1. Economizer: Economizers utilise the heat that remains from the combustion gases to warm the water. Before entering the boiler, the feed water is routed through the economizer. An economizer is a device used to improve boiler efficiency.
2. Centrifugal Pump: The Lamont boiler uses forced convection. To circulate water within the boiler, a centrifugal pump is employed. This pump is powered by a steam turbine. The boiler provides the steam needed by the turbine.
3. Evaporator Tube: The evaporator tube, also known as the water tube, is located near the boiler wall, increasing the heating surface of the boiler. This also pertains to the positive and negative aspects of the furnace and other machinery. These tubes' main purpose is to convert water into steam. This also helps to cool the furnace wall.
4. Grate: The grate is the area in the boiler where the fuel is burnt. It is located on the furnace's bottom side.
5. Furnace: The vertical furnace is utilized in the Lamont boiler. The furnace's main function is to burn fuel.
6. Super heater: The evaporator tube produces saturated steam. It may lead to corrosion if used directly in the steam turbine. As a result, the saturated steam is directed to the top heater, known as the super heater, where it may be heated to a higher temperature.
7. Water Steam Separator Drum: Outside of the boiler is the steam separator. The water steam separator drum separates the steam and transmits the water from the evaporator tube's water and steam combination, which is then sent to the super heater. The remained water is sent back to the economizer.
8. Air Preheater: The air pre-heater warms the air before it enters the boiler. An air preheater is a device that improves boiler efficiency.

➢ Working Principle of Lamont Boiler

The Lamont boiler operates on the basis of forced circulation of water within the boiler using a centrifugal pump. It depends on the pump for its operation. The centrifugal pump pumps the steam and water combination through the boiler's small diameter tubes. The water is forced into the economizer by a feed pump, where the temperature rises.

A centrifugal pump powered by the steam turbine forces this water into the evaporator tube. Inside the tube, a combination of saturated steam and water is created. This combination is transferred to the steam separator drum, which is located outside the boiler. The steam from the separator is sent to the super heater, where it is converted from saturated steam to superheated steam. The water is sent back to the economizer, where it passes through the evaporator tubes once again. The air from the air preheater enters the boiler and burns the fuel. The flue gases heat the evaporator tube before passing through the super heater. These gases combine to produce the super heater, which preheats the air before expelling it into the environment. This boiler has a working pressure of more than 170 bar and a steam production capability of around 50000 kg/hour at a temperature of 773 K.

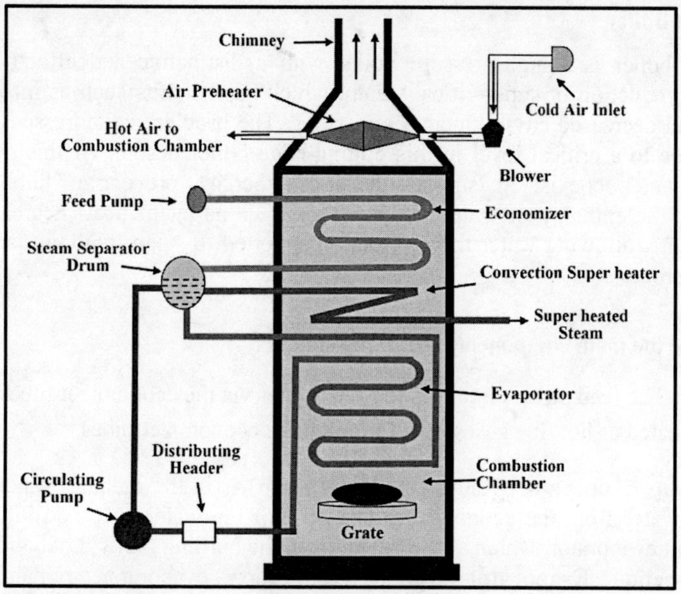

Fig. 3.5 Lamont Boiler

➢ Advantages of Lamont Boiler

- The Lamont boiler can produce a large volume of steam.
- It is simple to get start.
- The Lamont boiler's construction concept is simple yet effective.
- It has a high heat transmission rate and may be reassembled with the natural circulation boiler.
- Its design is customizable.

➢ Disadvantages of Lamont Boiler

- The creation of bubbles on the surface of the tube lowers the rate of heat transmission. As a result, there is a little issue with the overall quantity of steam generated.
- Expensive maintenance due to complex design.
- High construction expenses.
- Reduced efficiency and longer starting time.
- High pressure and temperature require specific preventative measures.
- Complex control system requires the use of skilled operators.
- Limited capacity for large-scale applications.
- Operates only with high-quality water.

3.6.2 Benson Boiler

The Benson boiler is a high-pressure boiler with a distinctive and effective design that enables the production of supercritical steam, which has benefits such as increased thermal efficiency and decreased environmental emissions. The problem is addressed by raising the boiler pressure to a critical level in this cutting-edge boiler design. At this pressure levels, steam and water coexist at similar densities, thereby preventing bubble formation. Additionally, the enthalpy of evaporation (also known as latent heat) decreases to zero at critical pressure, allowing water to be directly converted to steam without using the typical evaporation process.

Following are the main components of a Benson boiler:

Economizer: The feed pump circulates the feed water via the economizer tubes. The feed water is preheated as hot flue gases travel through the economizer tubes.

- ✓ **Radiant Evaporator:** Radiant parallel tube parts in the radiant evaporator carry the feed water from the economizer. The bulk of the water is turned into steam in the radiant evaporator, which absorbs heat from the burning fuel via radiation.
- ✓ **Convection Evaporator:** In the convection evaporator, remaining water is evaporated while being heated by convection from the hot gases. As a result, 210 kg/sq. cm of saturated high-pressure steam is generated.
- ✓ **Convection Super heater:** Now, the saturated steam flows through a convection super heater, where it is heated to 650°C. Convection evaporator, radiant evaporator, and convection super heater are all positioned in the flue gas route.
- ✓ **Steam Outlet:** The steam outlet is used to provide the superheated steam to the steam turbine.

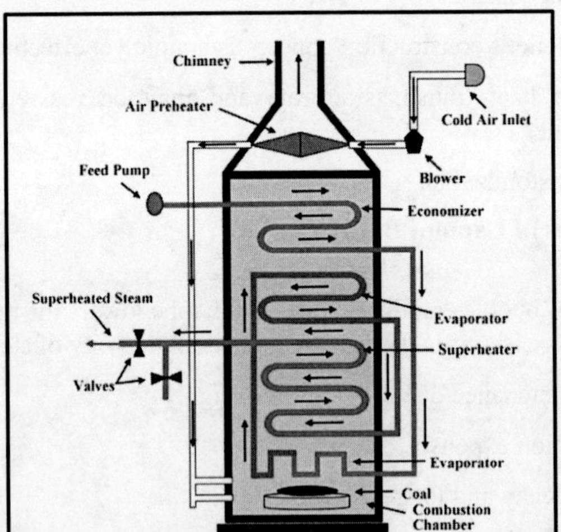

Fig. 3.6 Benson Boiler

- ➢ **Working of Benson Boiler**

It operates on water pressure that is raised to supercritical pressure, or pressure that is higher than the critical pressure of 225 bar. Water turns straight into steam without boiling when its pressure is raised to the supercritical level, when it has a latent heat of zero. Additionally, this stops bubbles from forming on the tube's surface.

The feed pump in a Benson Boiler raises water pressure to supercritical pressure before it reaches the economizer. The water travels from an economizer to the radiant heater. In this instance, the water is heated by radiation and partially turns into steam.

It almost reaches the supercritical temperature now. From that, the steam and water mixture flows into the convective evaporator, where it transforms into steam and may slightly overheated. The desired superheated steam is then obtained by passing it through the super heater at the end. Subsequently, turbines or engines utilize this superheated steam to generate energy.

➢ **Advantages**
- ✓ It has no drum and is lighter. As a result, it weighs less than other boilers.
- ✓ It starts up faster than conventional boilers.
- ✓ It has a thermal efficiency of up to 90%.

➢ **Disadvantages**
- ✓ Supercritical boilers require constant monitoring from the controller to prevent explosions.
- ✓ It is challenging to control the boiler under varying load.

3.7 Mountings

Steam boiler mountings refer to a collection of fittings that are affixed to the steam boiler in order to guarantee its effective and safe operation. Boiler mountings encompass several components such as the water level indication, safety valve, pressure gauge, and others. It is imperative to acknowledge that the operation of a boiler is inherently unsafe in the absence of its mountings. The installation of these fittings on the steam boiler is crucial for ensuring the safe and efficient functioning of the boiler system.

Some of the primary mountings found in the steam boiler are:

- **Water Level Indicator**

A water level indicator is positioned externally on the steam boiler shell in order to determine the water level within the boiler with the help of a transparent tube. The safety equipment in concern serves with the goal of ensuring the secure functioning of the steam boiler.

The Water-tube indication comprises a vertically oriented hard glass tube, equipped with two gunmetal tubes denoted as A and B. Tube A and B connect the steam area and water space of the boiler to the glass tube, respectively.

Tube A is equipped with a valve, referred to as a steam valve, whereas tube B is outfitted with a separate valve known as a water valve. Furthermore, the water level indication is equipped with a third valve known as the drain valve. The drain valve facilitates the regular removal of water and condensed steam from the gunmetal tube A.

- **Pressure Gauge**

The steam pressure in the boiler is shown by a pressure gauge. Typically, it is attached to the top front portion of the steam boiler.

The pressure gauge is comprised of a circular spring tube denoted as A. One terminal of the tube is sealed and connected to a linkage, whilst the other terminal is attached to a hollow block. The closed end of the tube is connected to the toothed sector, which is hinged, through the connection. The toothed sector gears are equipped with a pinion that is responsible for carrying a pointer. The dial is fitted with a pointer that serves the purpose of indicating the pressure units.

- **Safety Valve**

The safety valve is an essential safety device installed on the shell of a steam boiler, serving to protect the boiler from excessive pressure and maintain its operational safety. The primary function of this mechanism is to reduce the possibility of explosions resulting from the formation of excessive internal steam pressure inside the boiler system.

The safety valve is a device designed to prevent the steam boiler pressure from exceeding its prescribed operating pressure. The automatic opening of the system occurs when the pressure within the steam boiler exceeds the specified limit for normal operational conditions. Consequently, it permits the discharge of surplus steam into the atmosphere until the pressure is restored to its standard level. The use of a safety valve serves to protect the steam boiler by reducing any harm caused by an excessive formation of steam pressure. The primary types of safety valves include deadweight safety valve, lever safety valve and spring-loaded safety valve.

- **Steam Stop Valve**

The steam stop valve is responsible for regulating the flow of steam originating from the steam boiler, and also has the ability to cut off the flow entirely when needed. The steam stop valve performs the purpose of regulating the flow of steam inside the boiler system.

The valves affixed to the steam boilers, which provide a 90° alteration in the direction of steam flow, are commonly referred to as junction valves. On the other hand, valves that are installed within pipelines to enable the unidirectional flow of steam are referred to as stop valves.

- **Blow-Off Valve**

The blow-off valve serves the purpose of periodically releasing accumulated sediments from the bottom of the steam boiler during its operational phase, as well as facilitating the drainage of the steam boiler for the purposes of cleaning or inspection.

When the blow-off valve is activated, the water, which is pressurized by steam, escapes at a high velocity, resulting in the discharge of sediments.

- **Feed Check Valve**

In order to restore the water level in the steam boiler to its designated capacity, an excess amount of water is introduced. This is referred to as feed water. The elevated pressure within the boiler necessitates the utilisation of a pump to augment the feed water pressure prior to its introduction into the steam boiler. The feedwater, which is pressurised, is introduced into the boiler system via the feed check valve.

The feed check valve is responsible for regulating the water flow originating from the feed pump, while also serving as a barrier against the reverse flow of water from the boiler system back to the pump. Hence, the feed check valve is positioned at the termination of the delivery pipe of the feed pump, namely at the boiler end.

- **Fusible Plug**

The fusible plug is comprised of a low melting point alloy, often composed of tin or lead. The fusible plug serves the purpose of extinguishing the fire within the steam boiler furnace when the water level decreases below the established safety threshold. This action effectively reduces the risk of explosion that might arise from excessive heating of the tubes and shell. The component is installed at the highest point of the furnace or the combustion chamber.

3.8 Boiler Accessories

Boiler accessories play a crucial role in ensuring proper functionality and enhancing the efficiency of steam power plants. The boiler accessories are not directly installed with the boiler.

- **Economizer**

Significant quantities of thermal energy can be found inside the combustion gases that are discharged from the boiler. Hence, it is essential to maximize heat recovery from the gases prior to their release into the chimney. The economizer facilitates heat recovery from the exhaust gases through the process of heating the feed water. The economizer is installed in the pathway of the gases. By reducing fuel use, the overall efficiency of the boiler is enhanced.

- **Air Pre-heater**

The air preheater is an additional part of a boiler system that facilitates the recovery of thermal energy from the exhaust gases, therefore raising the temperature of the air supplied to the boiler furnace. The introduction of preheated air into the furnace helps to increase the temperature within the furnace, increasing the combustion process of the fuel. Consequently, the thermal efficiency of the plant will be enhanced.
The benefits of air preheating include:

- An increase in the rate of steam production
- Low-grade fuels can be used
- Combustion is improved with less soot, smoke, and ash produced

- **Superheater**

Within the context of boilers, a superheater component fulfils the purpose of raising the temperature of steam above its saturation point. The flue gases cross a series of tubes positioned along their path, whereby the dry, saturated steam generated by the boiler undergoes additional heating from the high-temperature gas, resulting in a temperature nearing saturation.

- **Feed Pump**

To force high-pressure feed water into the boiler, an accessory called a feed pump is required. Common types of pumps include rotational pumps and rotating pistons. The reciprocating pumps are directly driven by being connected to the steam engine. Rotary pumps are commonly powered by either electric or steam turbines.

3.9 Steam Separators

A steam separator, also known as a moisture separator or steam dryer, is a mechanical apparatus designed to effectively separate water droplets from steam. The most elementary form of steam separator is represented by the steam dome found on a steam locomotive. Stationary boilers and nuclear reactors sometimes have intricate mechanisms that induce a rotational motion in the steam, resulting in the expulsion of water droplets by centrifugal force, which are subsequently collected. Steam separators necessitate the use of steam traps in order to effectively gather the water droplets that are eliminated.
It is important to remove water droplets from steam because:

- The thermal efficiency of all engines is reduced by the presence of damp steam.
- In the context of piston engines, the presence of water within the cylinders can lead to a phenomenon known as hydraulic lock, resulting in detrimental effects on the engine. Additionally, this water intrusion can also have adverse consequences for the internal components of the turbine, particularly the turbine blades, by causing impingement and erosion.
- In various industrial machinery that utilise steam, the presence of water in the piping system can lead to the occurrence of steam hammer. This phenomenon is characterised by the accumulation of water within the pipe, which subsequently obstructs the flow of steam. As the steam continues to flow through the pipe, it accelerates the water, eventually reaching a sharp bend. At this point, the water's momentum causes a sudden and severe failure of the pipe, resulting in catastrophic consequences.

The classification of steam separators encompasses three fundamental categories: baffle, centrifugal, and mechanical coalescence. Frequently, the use of a dual combination of different separator types is employed in order to get enhanced efficiency.

✓ **Baffle Separator**

In terms of practicality and design, this kind of moisture separator is simple. The redirection of steam is facilitated by the presence of one or more internal baffles within the separator. The alteration of steam flow direction is the underlying cause for the elimination of the more substantial condensate droplets from the steam. A drainage apparatus, such as a steam trap station or control valve, facilitates the expulsion of dense condensate. Although the baffle separator may not possess the same level of efficiency as more modern counterparts, it remains a widely utilised option in straightforward applications.

✓ **Centrifugal Separator**

This particular method of condensate removal utilises centrifugal force as a means to extract the entrained condensate from the steam. To ensure proper sizing of a centrifugal separator, it is crucial to get accurate measurements of both pressure and steam flows. In order to achieve optimal separation performance, it is imperative to ensure that the centrifugal motion remains unimpaired.

✓ **Mechanical Coalescing/Centrifugal Separator**

The process of moisture removal in this particular separator occurs through a two-stage mechanism. During the initial phase, the steam undergoes a process of coalescence, resulting in the formation of bigger droplets upon its entry into a stainless steel mesh. In the subsequent phase, the aforementioned bigger droplets are afterwards eliminated by the use of a centrifugal component. The aforementioned technology has a remarkably high effectiveness in removing moisture. Typically, exhibiting an efficiency rate of around 99.9%.

In some systems, the presence of a steam separator is necessary in order to effectively manage the moisture content. Failure to incorporate a steam separator might result in the emergence of various operational issues.

- A decrease in the operational effectiveness of the system
- Breakdown of the basic elements of the system
- Frictional resistance during fluid flow
- Dysfunctions in flow metres and control valves
- Decrease in the overall productivity of the system

3.10 Steam traps

Steam traps are employed for the purpose of preventing the entry of condensate, air, and other undesirable contaminants into the steam supply. By engaging in this practise, individuals safeguard terminal operations and preserve the effectiveness of the system. Steam traps do not effectively capture or retain steam. In contrast, they effectively capture and eliminate all non-steam substances from the system while maintaining steam pressure intact. Undesirable condensate, gases, and various contaminants are permitted to accumulate inside a reservoir, thereafter being expelled periodically to maintain the purity of the steam. In the absence of a steam trap, the uncontrolled circulation of condensate and undesired air would occur within the steam system. The proposed action would not only result in a decrease in operational efficiency, but it would also lead to detrimental effects such as damage and corrosion inside the steam-dependent processes.

Different kinds of steam traps are available, based on their purpose and location in the boiler system

- Balanced Pressure Trap

Balanced pressure traps, sometimes referred to as liquid type traps, employ a liquid medium for the regulation of the valve within the steam trap. The state of the liquid within the valve is dependent upon the relationship between pressure and temperature, whereby it may undergo boiling or condensation, thus resulting in the activation or deactivation of the trap mechanism.

- *Thermostatic Traps*

These traps, which are employed in non-sensitive sub-cooling environments, depend on temperature as the triggering mechanism for their purge cycle.

- *Thermodynamic Traps*

Use flash steam to open and close the valve.

- *Bimetallic Traps*

The material has exceptional durability, although demonstrates permeability at low pressure conditions and experiences obstruction under high pressures and temperatures.

- *Float Traps*

Float traps, as its name suggests, employ a valve that is linked to a float. The valve is triggered to open if the condensate exceeds a specific threshold.

- *Inverted Bucket Traps*

The proposed method involves employing continuous increase of steam pressure to lift an inverted, bucket-shaped cylinder. Subsequently, this rise triggers the activation of a valve, facilitating the discharge of undesired air and condensed water vapour.

3.11 Difference between Steam Separator and Steam trap

Steam separator and steam trap are almost same in function. Both used to avoid the problem caused by water residue in the steam.

Steam Separator	**Steam trap**
The steam separator is a mechanical apparatus designed to effectively separate water particles from steam.	Steam traps are devices that effectively separate condensate and non-condensate gases, such as air and CO_2, from steam without any steam leakage.
The purpose of this device is to enhance the dryness fraction, or quality, of steam prior to its introduction into a turbine or engine.	Their primary purpose is to prevent the buildup of condensate in steam ducts and pipes.
Typically, a chamber is utilised for the purpose of collecting condensate, accompanied by an automated drain valve that is triggered based on the predetermined water level within the chamber.	These valves operate using distinct mechanisms and are purely automated in nature.
Large surface areas that are exposed to an environment that results in significant heat loss. Certain individuals require an insulating jacket in order to mitigate thermal energy dissipation.	The magnitude of heat loss is significantly lower in comparison to steam separators.
The primary categorizations are the baffle separator, centrifugal separator, and mechanical coalescing separator.	These traps may be categorised into three types: thermostatic trap, mechanical trap (also known as density trap), and thermodynamic trap.

3.12 Equivalent evaporation

The concept of equivalent evaporation is employed in the field of thermodynamics to simplify the comparison between the heat energy necessary for the evaporation of a specified quantity of water at a given temperature and the energy output resulting from the combustion of a particular fuel. The previously mentioned concept holds significant importance within the field of steam power production, as it serves as a means to evaluate the efficacy of steam engines and boilers. In many applications such as steam power plants, where steam is utilised for the purpose of generating mechanical work or electricity, the process involves heating water to produce steam. This steam is subsequently employed to drive turbines or carry out other designated functions. The optimisation of energy conversion relies heavily on the efficiency of this process.

The concept of equivalent evaporation allows a direct comparison between the thermal energy generated by the combustion of a certain fuel and the thermal energy necessary to convert water into steam. This comparative analysis is particularly valuable in scenarios when many fuel options are being evaluated for their suitability in steam boilers, including coal, oil, and natural gas. Through the process of computing the comparable evaporation for each fuel, engineers are able to determine the relative efficiency of different fuels in generating steam based on a specific quantity of heat energy input. The significance of the idea of equivalent evaporation is in its ability to aid in the selection of the optimal fuel for a given steam production application, hence enhancing energy utilisation and cost-effectiveness. Additionally, it facilitates the process of designing and optimising steam boilers and engines in order to get enhanced efficiency and overall performance.

In basic terms, equivalent evaporation serves as a metric that enables the evaluation of the thermal energy generated by fuel combustion in comparison with the energy needed for water vaporisation. This metric plays an essential part in the design and functioning of steam-driven systems utilised in power generation and industrial applications.

The term "enthalpy of evaporation" refers to the mass of water that is fully saturated at 100°C and is converted into dry, saturated steam at the same temperature, using heat at a rate equivalent to that which would have been employed in the real operational circumstances.

If H = Total heat of steam at given working pressure in kJ/kg

H_w = Total heat of feed water in kJ/kg

L = Latent heat of steam at atmospheric pressure = 2257 kJ/kg

W_a = Weight of steam produced per hour at given working pressure per kg of fuel

W_e = Equivalent evaporation in kg per kg of fuel

Then as per the above definition, $W_e \times L = W_a (H - H_w)$ (3.1)

Or $W_e = (W_a (H - H_w))/2257$ (3.2)

3.12.1 Factor of evaporation

The term "equivalent evaporation" refers to a specific number that, when multiplied by the actual amount of steam produced at a particular pressure from water at a specific temperature, gives the corresponding evaporation value.

Equivalent Evaporation = F x Actual Evaporation

The factor of evaporation can also be described as the ratio of equivalent evaporation to actual evaporation.

3.12.2 Boiler efficiency

The thermal efficiency refers to the proportion of heat effectively used in the production of steam during a period of time, relative to the heat provided by the fuel during the exact same period of time. Alternatively, it may be defined as the ratio of the heat utilised in generating a specific amount of steam to the heat supplied by the fuel consumed in the process of steam production.

$$Boiler\ Efficiency, \eta_b = \frac{W_a(H - H_{wi})}{W_f.C}\ kg \qquad (3.3)$$

Where C = calorific value of fuel in kJ/kg and W_f= weight of fuel per hour.

The efficiency of a boiler is constantly below 1 due to the loss of heat by the escape of hot gases into the atmosphere, as well as through conduction, convection, and radiation directly to the atmosphere.

3.12.3 Boiler horse power

The use of this word in the context of boilers lack justification due to its inability to generate mechanical work or power. However, it has been officially recognised by the American Society of Mechanical Engineers as a standard unit for quantifying boiler capacity.

One boiler horse power means evaporation of 15.653 kg of water per hour from and at 100°C into dry saturated steam at 100°C. Or a boiler generating 15.653 kg of equivalent evaporation per hour means developing one boiler H.P.

$$So, one\ Boiler\ H.P = \frac{Equivalent\ evaporation\ per\ hour}{15.653} \qquad (3.4)$$

Sample Problems

Numerical 3.1 A boiler evaporates 3.6 kg of water per kg of coal burnt into dry saturated steam at 10 bar pressure. The temperature of feed water is 32°C. Find the equivalent evaporation from and at 100°C as well as factor of evaporation.

Solution

Mass of steam produced, $m_s = 3.6\ kg/kg\ of\ coal$

Initial temperature of feed water, $t_w = 32\ ℃$

So, Enthalpy of feed water, $H_1 = m.c_p.t_w = 3.6 \times 4.187 \times 32 = 482.34\ kJ$

Enthalpy of dry saturated steam produced

$H_2 = m.\ h_g$

At 10 bar pressure, the value of h_g from steam tables is = $2776.2\ kJ/kg$

So, $H_2 = 3.6 \times 2776.2 = 9994.3\ kJ$

Hence the heat supplied to steam per kg of coal burnt in the boiler is

$H_2 - H_1 = 9994.3 - 482.3 = 9512\ kJ$

Equivalent Evaporation of Boiler;

$$= \frac{H_2 - H_1}{2257} = \frac{9512}{2257} = 4.kg/kg\ of\ coal\ burnt\ (\mathbf{Ans})$$

$$\text{Factor of Evaporation} = \frac{Equivalent\ Evaporation}{Mass\ of\ steam\ produced}$$

$$= \frac{4.2}{3.6} = 1.17$$

(Ans)

Numerical 3.2 The following readings were obtained during a boiler trial of 6-hour duration

Mean Steam Pressure = 12 bar

Mass of steam generated = 40000Kg

Mean dryness fraction = 0.85

Mean feed water temp = 30° C

Coal Used = 4000kg

Calorific value of coal = 33400 kJ/kg

Calculate: (1) Factor of equivalent evaporation

(2) Equivalent evaporation

(3) Efficiency of boiler

Solution

Mass of steam produced in 6 hours duration = 40000 Kg

Mass of coal burnt in 6 hours duration = 4000kg

Thus mass of steam produced per kg of coal

$$m_s = \frac{40000}{4000} = 10\ \text{kg/kg of coal burnt}$$

Now specific enthalpy of feed water at temperature 30°C

$h_1 = C_p t = 4.187 \times 30 = 125.6$ kJ/kg

Specific enthalpy of steam produced at 12 bar pressure and dryness fraction 0.85

$$h_2 = h_f + x h_{fg}$$

Now at 12 bar pressure value of h_f and h_{fg} from Steam Tables are

$h_f = 798.4$ kJ/kg

$h_{fg} = 1984.3$ kJ/kg

Putting in above equation

$h_2 = 798.4 + 0.85 \times 1984.3 = 2485$ kJ/kg

$$\text{Factor of Evaporation, F} = \frac{h_2 - h_1}{2257} = \frac{2485-125.6}{2257} = 1.045$$ **(Ans)**

Equivalent Evaporation = mass of steam x Factor of Evaporation

= 10 x 1.045 = 10.45 kg/kg of coal burnt **(Ans)**

Efficiency of boiler, $\eta = \dfrac{m_s(h_2 - h_1)}{C}$

$$= \frac{10(2485 - 125.6)}{33400} = 70.6\ \%$$

3.13 Draught systems

The term "boiler draught" may be described as the minimal difference in pressure between the external air and the gases present within a furnace or chimney at the grate level. This pressure differential facilitates the movement of air and hot flue gases through the boiler.

The use of draught is essential in facilitating the passage of air through the fuel bed or grate, hence assisting in the efficient burning of fuel. Additionally, it serves the purpose of releasing the byproducts of combustion, commonly referred to as flue gases, into the atmosphere subsequent to their transfer of heat to the water undergoing evaporation within the boiler. The presence of draught in a boiler system contributes to the velocity of flue gases, hence enhancing the heat transfer coefficient. The presence of draught in a boiler is a fundamental need, and it may be generated through several means. Generally, draught systems can be classified into two categories: Natural draught and Artificial draught.

- **Natural Draught**

The natural draught method utilises a chimney, as seen in the following image. The chimney is a vertical cylindrical structure made of masonry or reinforced concrete materials. The purpose of its construction is to enclose a vertical column of exhaust fumes in order to generate a draught. The emission of gases takes place at a height that effectively reduces the risk of atmospheric pollution. The phenomenon of draught is generated by the presence of a tall chimney, resulting from the disparity in temperature between the hot gases within the chimney and the colder ambient air outside the chimney.

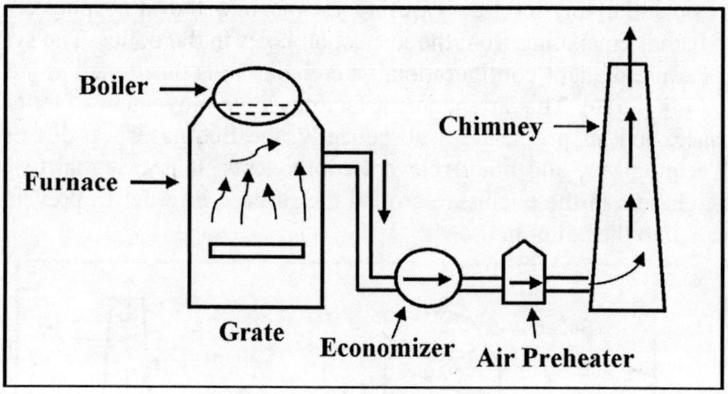

Fig. 3.7 Natural Draught

- **Artificial Draught**

In order to work well, the power plant needs a lot of water (300 mm of draught). It is necessary to use a different method, known as artificial draught, in order to meet these higher draught needs. Making artificial airflow is the job of the mechanical draught, which is often called "dan" (which comes from the word "fan"). When it comes to major power plants, automatic flow is usually the best option. The artificial draught can be further classified as:

- ✓ **Induced draught**

In this method, a fan or blower is placed at or very close to the chimney's base. Because of this placement, there is a partial vacuum in both the boiler and the flue tunnel. Because of this, the air and waste fumes are finally sucked through the boiler because the outside atmosphere is under a lot of pressure. Similarly to forced draught, any fan or blower can be used to make induced draught. A fan that creates negative pressure inside the boiler is used to keep it that way. This fan is placed before the chimney. One major problem with this system, though, is that air could get into the boiler if the seals aren't tight enough. This would mean that the system wouldn't work as well as it should, and money would be wasted.

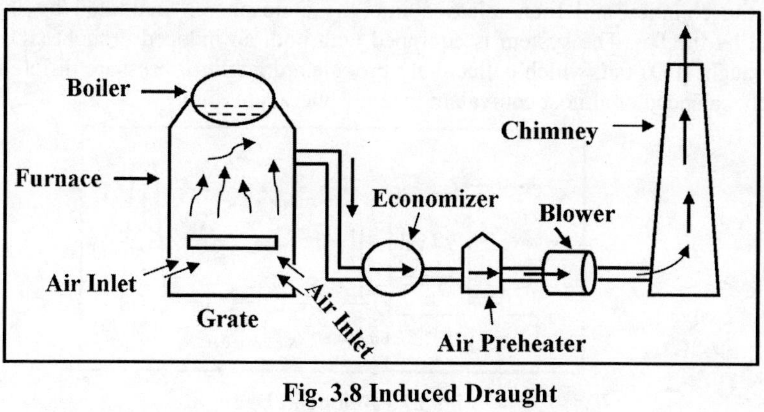

Fig. 3.8 Induced Draught

✓ **Forced draught**

The generation of positive pressure inside the boiler is facilitated by the use of a forced draught fan, which is positioned prior to the boiler. This fan enables the introduction of air into the boiler, resulting in the creation of positive pressure. However, a drawback of this system is the potential for hot flue gases to escape into the atmosphere, as well as the occurrence of flames emanating from the inspection holes in the boiler. The system described is a positive pressure draught configuration, whereby a fan is positioned at the bottom of the boiler, before to the grate. This arrangement facilitates the movement of external air via the fuel bed, furnace, and air pre-heater. Subsequently, the flue gases are directed through the flue channel, economizer, and other relevant components. It is essential to provide a high degree of air-tightness in the enclosure around the furnace in order to prevent any potential leakage of gases into the boiler house.

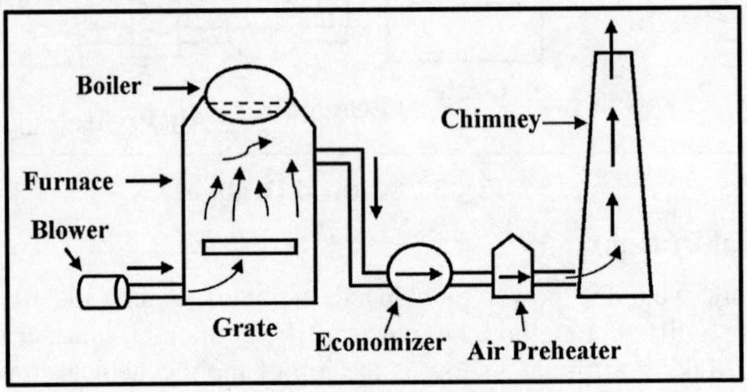

Fig. 3.9 Forced Draught

✓ **Balanced draught**

The system utilises both forced and induced draught mechanisms. Within this particular system, the forced draught fan effectively reduces the resistance encountered in the air pre-heater and grate. Conversely, the induced draught fan is responsible for compensating for the draught losses that occur throughout the boiler, economizer, and connected flue passages, among other components. The utilisation of balanced draught is very efficient and primarily employed in the operation of sizable boilers. The selection of the fan or blower used in a boiler system is contingent upon the specific fuel being burned and the type of boiler in operation. This may encompass a range of fan types, including radial or axial configurations, among others. Typically, it is powered by an electric motor. The air blower is installed within the burner assembly in the context of oil burners. Ultimately, the flue gases are directed towards the chimney and then released into the atmosphere, facilitated by the pressure generated by the fan. The system is equipped with both an induced draught (ID) fan and a forced draught (FD) fan, which collectively provide a draught or pressure differential that is effectively balanced or almost equivalent to atmospheric pressure.

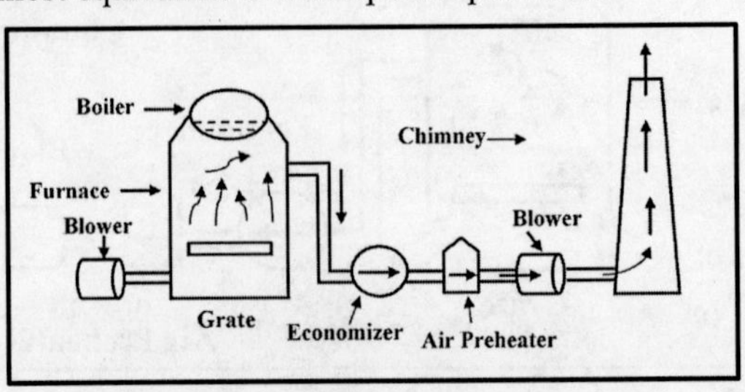

Fig. 3.10 Balanced Draught

3.14 Comparison between Forced Draught and Induced Draught

Forced Draught	Induced Draught
1) In the case of an oil-fired boiler, a fan or blower is positioned either before the grate or within the oil burner assembly.	The fan or blower is positioned downstream of the grate or flue passageways and upstream of the chimney.
2) The pressure within the flue tunnels is marginally higher than the surrounding air pressure.	The pressure within the flue tunnels is marginally lower than the surrounding air pressure.
3) As a consequence of the aforementioned factor, there exists a significant risk of fire occurrence, potentially resulting in harm to individuals, in the event of any leakage within the boiler flue route.	As a result of the aforementioned factor, there is no potential risk of fire emanating from the boiler flue route in the event of a leakage. Therefore, induced draught is considered to be a safer option.
4) The introduction of fresh air into the combustion chamber facilitates the combustion of fuel and the generation of high-temperature combustion gases. As a result of this force, the high-temperature flue gases are propelled through the flue passageways within the boiler.	The system effectively extracts the hot flue gases generated in the combustion chamber and directs them through flue passageways, afterwards transferring them to the economizer, air preheater, and chimney. As a result of this suction phenomenon, a flow of fresh air is drawn into the combustion chamber.
5) Forced draught fans consume less power due to its operation with cold and dense air, which possesses a lower volume per unit mass.	The induced draught fan necessitates a greater amount of power due to its responsibility of managing high-temperature flue gases, which possess a higher volume per unit mass.
6) The flow of flue gases through the boiler exhibits a higher degree of uniformity.	The uniformity of flue gas flow is reduced.

UNIT SUMMARY

Based on the contents covered in this unit you should remember that:

- A 'boiler' is a sealed container where steam is generated from water through the burning of fuel.
- Fire tube boilers include an arrangement where the hot gases are contained inside the tubes, while the tubes are surrounded by water.
- Water tube boilers have a configuration where the water is contained inside the tubes, while the surrounding area is occupied by hot gasses.
- Modern high pressure boilers used for power production have steam capacity ranging from 30 to 650 tonnes/h and higher. They may operate at pressures up to 160 bar and have a minimum steam temperature of around 540°C.
- Boiler mountings refer to a number of attachments and devices that are essential for the functioning and safety of a boiler.
- Boiler accessories are components that are fitted in order to enhance the overall efficiency of a steam power plant.
- The evaporation rate refers to the amount of water that is converted into steam during a one-hour period.

- The factor of evaporation is the ratio of the actual heat absorbed above the temperature of the feed-water, used for converting it into steam, divided by the latent heat of steam at atmospheric pressure.
- Draught refers to the small pressure variation that is deliberately maintained in the boiler in order to remove the combustion by products from the combustion chamber.
- Natural draughts generated by chimneys or artificial draughts generated by steam jets or mechanical devices.

Exercises

Multiple Choice Questions

1. **Which of the following is a primary classification of boilers based on their operating pressure?**

   A) Fire-tube and water-tube boilers

   B) Horizontal and vertical boilers

   C) Single-phase and three-phase boilers

   D) Low-pressure and high-pressure boilers

2. **What classification of boilers is based on the location of the combustion chamber relative to the heat exchange surfaces?**

   A) Fire-tube and water-tube boilers

   B) Natural and forced draught boilers

   C) Single-flame and multi-flame boilers

   D) Steam and hot water boilers

3. **Which of the following is a primary function of a safety valve in a boiler?**

   A) To control the water level in the boiler

   B) To release excess steam and prevent overpressure

   C) To measure the steam temperature

   D) To supply water to the boiler

4. **What does a water level indicator do in a boiler system?**

   A) Measures the temperature of the steam

   B) Controls the fuel flow into the boiler

   C) Shows the current level of water inside the boiler

   D) Regulates the air supply for combustion

5. **What is the primary purpose of a draught system in a boiler?**

   A) To regulate the water level in the boiler

   B) To control the combustion air and remove flue gases

   C) To monitor steam pressure

   D) To measure feedwater temperature

6. **Which type of draught system uses a fan to induce air into the combustion chamber?**

   A) Natural draught

   B) Induced draught

   C) Forced draught

   D) Balanced draught

7. **In a balanced draught system, which components work together to maintain the required draught?**

   A) Induced draught fan and chimney

   B) Forced draught fan and chimney

   C) Induced draught fan and forced draught fan

   D) Steam generator and economizer

8. **What is the primary measure of boiler efficiency?**

   A) Energy output relative to energy input

   B) Heat transfer

   C) Fuel consumption

   D) Steam temperature

9. **Which of the following factors can decrease boiler efficiency?**

   A) Proper maintenance

   B) Scale and soot buildup

   C) High-quality fuel

   D) Adequate combustion air

10. **What is the purpose of a boiler economizer?**

    A) To increase steam pressure

    B) To recover waste heat from flue gases

    C) To improve fuel quality

    D) To control water levels

Answer: B) To recover waste heat from flue gases

Answer Key:

1 D	2 A	3 B	4 C	5 B
6 C	7 C	8 A	9 B	10 B

Short and Long Answer Type Questions

- What does the term steam generator mean?
- Explain the function and uses of steam boilers.
- What is the basic difference between water tube and fire tube boilers?

- What is "thermal efficiency" in the context of a boiler?
- How can regular maintenance improve boiler efficiency?
- What are the purposes of carrying out boiler trial?
- Differentiate between natural and artificial draught.
- Discuss the design, operation, and applications of high-pressure boilers, including the benefits and challenges associated with their use.
- Explain the different types of boilers used in industrial applications, including their design, operation, advantages, and limitations.
- Explain the various mountings and accessories used in boilers, detailing their functions, importance, and how they contribute to the efficient and safe operation of the boiler.
- Discuss the factors that affect boiler efficiency and the methods to improve it.

Know More

- Now a day, steam generators comprises of advanced control systems, such as distributed control systems (DCS) and programmable logic controllers (PLC), that controls their operations. These advanced systems use sensors and actuators for improving the performance and efficiency.
- Computational fluid dynamics simulations are often used in the design and optimization of modern steam generators. Computational Fluid Dynamics (CFD) enables engineers to simulate and examine the movement of gases and liquids within the boiler, with the aim of enhancing combustion efficiency, heat transfer, and overall operational effectiveness.
- Steam generators are subject to significant issues related to fouling and corrosion. Fouling is waste deposition on surfaces involved in heat exchange which in result decreases efficiency. The chemical processes that take place between water, steam, and the materials used in the boiler causes corrosion. These problems can be overcome by using technologically advanced materials and regular inspections.
- Steam generators work under high pressure and temperature, requiring the use of durable materials and a robust structure.
- The efficiency of a steam generator is impacted by variables such as heat transfer efficiency, fuel type, and combustion conditions.
- The efficiency of a steam generator is affected by variables such as heat transfer efficiency, fuel composition, and combustion conditions.

References and Suggested Readings

- Thermal Engineering by B. R. Sarkar (McGraw-Hill Education (India) Pvt Limited).
- Thermal Engineering by R.K. Rajput (Laxmi Publications (P) Ltd).
- A Textbook of Thermal Engineering by R.S. Khurmi & J.K. Gupta (S. Chand & Company Ltd).
- Thermal Engineering by Mahesh. M. Rathore (McGraw-Hill Education (India) Pvt Limited).
- Thermal Engineering by Sadhu Singh and Sukumar Pati (Pearson India Education Services Pvt. Ltd,).

Dynamic QR Code for Further Reading

1. Lectures of the NPTEL Course on "Types of Boiler, Different Cycles in Boiler Operation, Boiler attachment" by Dr P.K. Mondal, Department of Mechanical Engineering, Indian Institute of Technology Guwahati, India.

2. Lectures of the NPTEL Course on "Cochran Boiler Operation, Boiler attachment" by Dr P.K. Mondal, Department of Mechanical Engineering, Indian Institute of Technology Guwahati, India.

3. Lectures on "Boiler Mountings and Accessories" by Prof. Ravi Kumar, Department of Mechanical & Industrial Engineering, Indian Institute of Technology Roorkee, India.

4. Lectures on "Draught" by RKB EduHub.

4 Steam Nozzles

UNIT SPECIFICS

This unit presents information related to the following topics:

- Flow of steam through nozzles.
- Velocity of steam at the exit of nozzle in terms of heat drop using analytical method and Mollier chart.
- Discharge of steam through nozzles and critical pressure ratio.
- Methods of calculation of cross-sectional areas at throat and exit for maximum discharge.
- Derivations on effect of friction in nozzles and Super saturated flow in nozzles and Working steam jet injector

This unit provides in-depth knowledge about the various types of nozzles and the flow of steam through them. The content provides detailed derivations for measuring the steam velocity at the nozzle's exit in terms of heat drop. The discharge of steam through nozzles is an important topic in the domains of thermodynamics, especially in relation to steam turbines and other systems that involve steam flow. The unit presents the related derivations in a straightforward manner, facilitating readers' understanding of the discharge phenomenon. The calculations required for obtaining cross-sectional areas at throat and exit for maximum discharge are also included. The unit concludes with the corresponding numerical data.

The unit concludes with a concise review of the key concepts. Additionally, the unit includes a significant portion of multiple-choice and descriptive questions. The unit includes a list of readings and sources for students to utilize for practice. It is important to mention that QR codes have been included in several parts to give more information on various areas of interest. The readers can scan these codes to obtain pertinent and corroborating information. The unit emphasizes the inclusion of video resources and QR codes to provide more information on numerous areas of interest. Mobile phones can access and view these resources by browsing or scanning QR codes.

RATIONALE

As the steam expands from a higher pressure to a lower pressure, the steam nozzle, a passage with varying cross-section, converts a portion of its enthalpy into kinetic energy. Therefore, the purpose of a nozzle is to enhance the steam's velocity. This unit will help the readers to understand the various factors that affect the velocity and discharge of steam through a nozzle. Frictional losses occur due to the interaction between the steam and the nozzle walls, which affects the nozzle's efficiency. This unit will help learners gain understanding of these repercussions, which are essential for precise performance predictions and design improvements. The detailed explanation of the steam injector is also provided for the readers, as it is a critical component in various industrial and power generation applications. Its importance stems from its ability to perform several essential functions efficiently and reliably.

PRE-REQUISITES

Before reading this unit, the students are advised to revisit the following:
Basic Mechanical Engineering (MEPC102)

UNIT OUTCOMES

After studying this unit students will be able to:

U4-O1: Understand the flow of steam through nozzles.

U4-O2: Learn about the velocity of steam at the exit of nozzle in terms of heat drop.

U4-O3: Calculate Discharge of steam through nozzles and critical pressure ratio.

U4-O4: Learn about the Methods of calculation of cross-sectional areas at throat and exit for maximum discharge.

U4-O5: Understand the effect of friction in nozzles and Super saturated flow in nozzles and working steam jet injector.

Unit 4 Outcomes	**EXPECTED MAPPING WITH COURSE OUTCOMES** (1- Weak Correlation; 2- Medium Correlation; 3- Strong Correlation)				
	CO-1	**CO-2**	**CO-3**	**CO-4**	**CO-5**
U4-O1	-	-	-	3	-
U4-O2	-	-	-	3	-
U4-O3	-	-	-	3	-
U4-O4	-	-	-	3	-
U4-O5	-	-	-	3	-

4.0 Nozzle

A nozzle is a device that alters the cross-sectional area of a fluid in order to perform the process of converting the pressure energy of the fluid into kinetic energy. At any given cross-section of the nozzle, the mass flow rate of the steam doesn't change at all. As fluid flows through the nozzle, its velocity increases while its pressure falls. Among the many uses for nozzles are jet engines, gas turbines, steam turbines, flow measuring equipment, internal combustion engine fuel injection and carburetion systems, spray paints, and many more. A decrease in steam's enthalpy allows the steam jet to escape the nozzle at greater velocity.

4.1 Classification of Nozzles

The steam nozzle can be classified in three categories:

a) Convergent nozzle
b) Divergent nozzle
c) Convergent-Divergent nozzle

A nozzle can be termed as convergent when the cross sectional area decreases from entrance to exit uniformly. The convergent nozzle (Fig. 1(a)) is preferably used for non-compressible fluids when the back pressure is greater than or equal to critical pressure ratio. A divergent nozzle (Fig. 1(b)) is a nozzle whose cross sectional area increases from entrance to exit uniformly. This nozzle is used when the back pressure is less than critical pressure ratio. The convergent-divergent nozzle, as seen in Fig. 1(c), can be defined as nozzle having a cross-sectional area that reduces from the entry to the throat and then rises from the throat to the exit. The convergent divergent nozzles are used when the back pressure is lower than the critical pressure.

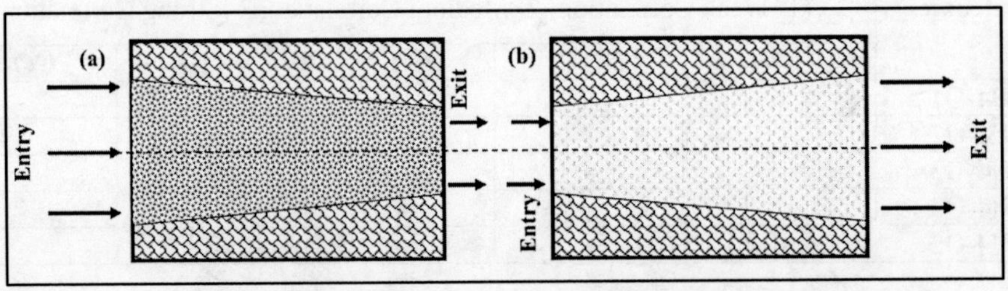

Fig. 4.1 (a) Convergent nozzle (b) Divergent nozzle

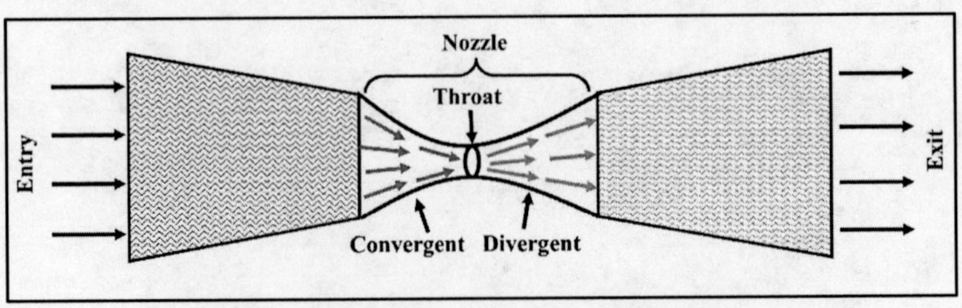

Fig. 4.1 (c) Convergent-Divergent nozzle

4.2 Flow of Steam through Nozzle

The steam flow through the nozzle may be regarded as an adiabatic flow because no heat is added or extracted during the expansion of steam in the nozzle. However, work is carried out through improving the kinetic energy of the steam. As the steam flows through the nozzle, it undergoes a reduction in temperature as well as pressure. The work done is equivalent to the adiabatic decrease in heat, which is in turn equal to the Rankine area.

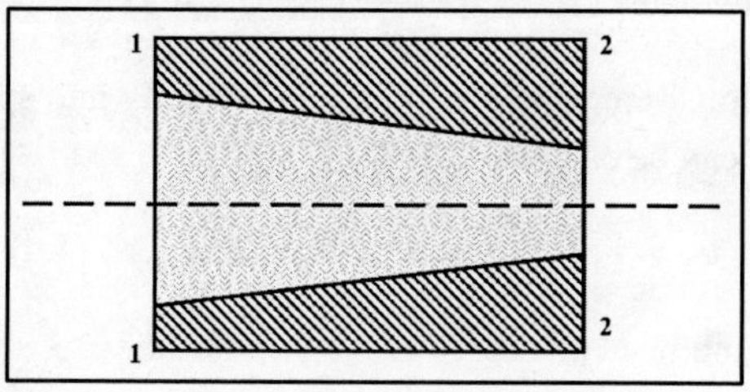

Fig. 4.2 Flow of steam through nozzle

Let us suppose a nozzle as shown in figure 4.2. Now, by applying energy equation at both the section for 1 kg/s flow, the relation becomes

$$\left(\frac{V_2^2-V_1^2}{2\times1000}\right) = h_1 - h_2 \quad \ldots\ldots\ldots\ldots\ldots\ldots (4.1)$$

Here h_1, h_2 and V_1, V_2 are the enthalpy (kJ/kg) and velocities (m/s) at section 1 and section 2 respectively.

From the equation 4.1, $V_2 = \sqrt{2000(h_1 - h_2) + V_1^2}$

At the entrance of the nozzle, the velocity of flow is considered as very low. Therefore, velocity at the entrance can be neglected and the equation becomes

$$V_2 = \sqrt{2000(h_1 - h_2)}$$

$$V_2 = 44.72\sqrt{(h_1 - h_2)} \text{ m/s}$$

The equation for the air velocity at exit can be written as

$$V_2 = 44.72\sqrt{C_p(T_1 - T_2) \times \eta_n} \quad \ldots\ldots\ldots\ldots\ldots\ldots (4.2)$$

Here C_p and η_n are specific heat at constant pressure and frictional loss, respectively.

4.3 Mass of Discharged Steam

The flow of the steam through the nozzle is isentropic flow which means the entropy of the flow is constant and it can be represented as $Pv^n = C$ where the value of n can be 1.135 and 1.3 for saturated steam and superheated steam respectively.

The work done at the entrance of nozzle will be: P_1v_1

The work done during expansion of steam in nozzle will be: $\frac{1}{n-1}(P_1v_1 - P_2v_2)$

The work done at the exit of nozzle will be: P_2v_2

The net work done will be, $W = P_1v_1 + {}^1/_{n-1}(P_1v_1 - P_2v_2) - P_2v_2$

$$W = ({}^n/_{n-1})(P_1v_1 - P_2v_2)$$

Now the equation becomes; $W = h_1 - h_2$

Mass of steam can be calculated as; $m = \frac{A \times V_2}{v_2}$ .... (4.3)

here A is the cross sectional area and V_2 and v_2 are velocity and specific volume at the exit section. The value of v_2 can be calculated as $P_1v_1^n = P_2v_2^n$;

$$v_2 = v_1 \times ({}^{P_2}/_{P_1})^{-1/n} \quad \text{.............. (4.4)}$$

Here v_1 is the specific volume at pressure P_1.

Now using equation 4.4, equation 4.3 becomes

$$m = \frac{A \times V_2}{v_2}$$

$$m = (A/v_2) \times \sqrt{2000(h_1 - h_2)} \quad \text{.................... (4.5)}$$

As $W = h_1 - h_2$ and $W = ({}^n/_{n-1})(P_1v_1 - P_2v_2)$; the equation 4.5 becomes

$$m = (A/v_2) \times \sqrt{2000 \times ({}^n/_{n-1})(P_1v_1 - P_2v_2)}$$

$$m = \frac{A}{(v_1 \times ({}^{P_2}/_{P_1})^{-1/n})} \times \sqrt{2000 \times ({}^n/_{n-1})P_1v_1\left(1 - \frac{P_2v_2}{P_1v_1}\right)}$$

$$m = \frac{A}{(v_1 \times ({}^{P_2}/_{P_1})^{-1/n})} \times \sqrt{2000 \times ({}^n/_{n-1})P_1v_1\left(1 - \left(\frac{P_2}{P_1}\right)\right)^{\frac{n-1}{n}}}$$

$$m = A \times \sqrt{2000 \times ({}^n/_{n-1})P_1/v_1\left(\left(\frac{P_2}{P_1}\right)^{\frac{2}{n}} - \left(\frac{P_2}{P_1}\right)^{\frac{n+1}{n}}\right)} \quad \text{................ (4.6)}$$

4.4 Discharge of steam through nozzles

In the convergent divergent nozzle, the pressure of the steam increases as it reaches the nozzle, but the velocity of the steam is almost insignificant. There is a decrease in the pressure of the steam as its velocity increases in the converging region, which is the portion that expands from the intake to the throat. Additionally, there is a decrease in the enthalpy, which is the total heat of the steam. The decrease in heat or enthalpy is turned into kinetic energy rather than being used for external effort. The pressure of the steam decreases further in the diverging region, which is the portion that extends from the throat to the outlet, while the velocity of the steam increases even further. Once again, there is a decrease in the enthalpy, also known as the total heat of steam, which converts into kinetic energy. The fact that the steam reaches the nozzle at a high pressure and with a minimum velocity will be an interesting aspect to keep in mind. However, the output of the nozzle is characterized by a high velocity and a low pressure. Back pressure refers to the pressure that is present when the steam is at its point of departure from the nozzle. In addition, even when the steam is flowing via a nozzle, it does not either provide or reject any heat. Because of this, it is regarded to be an isentropic flow, and the expansion that corresponds to it is also considered to be an isentropic expansion.

4.5 Maximum Discharge of Steam through Nozzle

It is possible to determine the maximum discharge from the nozzle by determining the specific throat pressure. This can be achieved by assuming that P_1 and P_2 represent the initial pressure and the pressure at the throat of the steam in N/m^2, and that v_1 and v_2 represent the volume of 1 kilogram of steam at pressures P_1 and P_2 respectively. The mass of the discharged steam was calculated as:

$$m = A \times \sqrt{2000 \times \left({}^{n}/_{n-1}\right) P_1/v_1 \left(\left(\frac{P_2}{P_1}\right)^{\frac{2}{n}} - \left(\frac{P_2}{P_1}\right)^{\frac{n+1}{n}}\right)}$$

The ratio of the steam pressure at the throat and initial pressure can provide the value of maximum discharge which can be achieved from the nozzle. The ratio of the steam pressure at the throat and initial pressure i.e. $\left(\frac{P_2}{P_1}\right)$ can be obtained through differentiation. So, by differentiating the part of equation which carry $\left(\frac{P_2}{P_1}\right)$ and equating with zero gives the following relation:

$$\frac{d}{d\left(\frac{P_2}{P_1}\right)}\left(\left(\frac{P_2}{P_1}\right)^{\frac{2}{n}} - \left(\frac{P_2}{P_1}\right)^{\frac{n+1}{n}}\right) = 0$$

$$\frac{2}{n}\left(\frac{P_2}{P_1}\right)^{\frac{2}{n}-1} - \frac{n+1}{n}\left(\frac{P_2}{P_1}\right)^{\frac{n+1}{n}-1} = 0$$

$$\frac{2}{n}\left(\frac{P_2}{P_1}\right)^{\frac{2-n}{n}} = \frac{n+1}{n}\left(\frac{P_2}{P_1}\right)^{\frac{n+1}{n}-1}$$

$$\frac{2}{n}\left(\frac{P_2}{P_1}\right)^{\frac{2-n}{n}} = \frac{n+1}{n}\left(\frac{P_2}{P_1}\right)^{\frac{1}{n}}$$

Now by putting $\left(\frac{P_2}{P_1}\right)$ terms on one side, the equation becomes

$$\frac{2}{n} \times \frac{n}{n+1} = \left(\frac{P_2}{P_1}\right)^{\frac{1}{n}} \left(\frac{P_2}{P_1}\right)^{-\frac{2-n}{n}}$$

$$\frac{2}{n+1} = \left(\frac{P_2}{P_1}\right)^{\frac{1}{n}-\frac{2-n}{n}}$$

$$\frac{2}{n+1} = \left(\frac{P_2}{P_1}\right)^{\frac{n-1}{n}}$$

$$\frac{P_2}{P_1} = \left(\frac{2}{n+1}\right)^{\frac{n}{n-1}} \quad \text{.................} (4.7)$$

Using this relation, the equation becomes

$$m = A \times \sqrt{2000 \times \left(\frac{n}{n-1}\right) P_1/v_1 \left(\left(\frac{2}{n+1}\right)^{\frac{2}{n-1}} - \left(\frac{2}{n+1}\right)^{\frac{n+1}{n-1}}\right)}$$

$$m = A \times \sqrt{2000 \times \left(\frac{n}{n-1}\right) P_1/v_1 \left(\frac{2}{n+1}\right)^{\frac{2}{n-1}} \left(\left(\frac{2}{n+1}\right)^{\frac{2}{n-1}-\frac{n+1}{n-1}} - 1\right)}$$

$$m = A \times \sqrt{2000 \times \left(\frac{n}{n-1}\right) P_1/v_1 \left(\frac{2}{n+1}\right)^{\frac{n+1}{n-1}} \left(\left(\frac{2}{n+1}\right)^{-1} - 1\right)}$$

$$m = A \times \sqrt{2000 \times \left(\frac{n}{n-1}\right) P_1/v_1 \left(\frac{2}{n+1}\right)^{\frac{n+1}{n-1}} \left(\left(\frac{n+1}{2}\right) - 1\right)}$$

$$m = A \times \sqrt{2000 \times \left(\frac{n}{n-1}\right) P_1/v_1 \left(\frac{2}{n+1}\right)^{\frac{n+1}{n-1}} \left(\frac{n-1}{2}\right)}$$

$$m_{max} = A \times \sqrt{1000\, n \times P_1/v_1 \left(\frac{2}{n+1}\right)^{\frac{n+1}{n-1}}} \quad \text{..................................} (4.8)$$

The equation 4.8 shows maximum discharge of steam through a nozzle.

4.6 Velocity of Steam at the Exit of Nozzle in terms of Heat Drop

Let us suppose that P_1 and P_2 represent the initial pressure and the pressure at the throat of the steam in N/m^2, and that v_1 and v_2 represent the volume of 1 kilogram of steam at pressures P_1 and P_2 respectively. Let V is the velocity of steam and A is the cross sectional area of nozzle at the throat. As shown in section 4.3, the flow of the steam through the nozzle is isentropic flow which means the entropy of the flow is constant and it can be represented as $Pv^n = C$ where the value of n can be 1.135 and 1.3 for saturated steam and superheated steam respectively. Referring section 4.3, the net work done will be

$$W = \left(\frac{n}{n-1}\right)(P_1 v_1 - P_2 v_2)$$

Now, the gain in kinetic energy $\left(\frac{1}{2}mV^2\right)$ will be equal to the net work done.

Therefore, $\frac{1}{2}mV^2$= Net work done (W)

$$\frac{1}{2}mV^2 = \left(\frac{n}{n-1}\right)(P_1v_1 - P_2v_2)$$

$$\frac{1}{2}mV^2 = \left(\frac{n}{n-1}\right)P_1v_1\left(1 - \frac{P_2v_2}{P_1v_1}\right)$$

As we know, $P_1v_1^n = P_1v_2^n$

$\therefore v_2 = v_1\left[\frac{P_1}{P_2}\right]^{\frac{1}{n}}$ or $\frac{v_2}{v_1} = \left[\frac{P_1}{P_2}\right]^{\frac{1}{n}}$

Now by using value of $\frac{v_2}{v_1}$; the equation becomes

$$\frac{1}{2}mV^2 = \left(\frac{n}{n-1}\right)P_1v_1\left(1 - \frac{P_2}{P_1}\left[\frac{P_1}{P_2}\right]^{\frac{1}{n}}\right)$$

$$\frac{1}{2}mV^2 = \left(\frac{n}{n-1}\right)P_1v_1\left(1 - \left[\frac{P_2}{P_1}\right]^{1-\frac{1}{n}}\right)$$

$$\frac{1}{2}V^2 = \left(\frac{n}{n-1}\right)P_1v_1\left(1 - \left[\frac{P_2}{P_1}\right]^{1-\frac{1}{n}}\right) \quad (\because \text{mass of steam is 1 kg})$$

$$V = \sqrt{2\left(\frac{n}{n-1}\right)P_1v_1\left(1 - \left[\frac{P_2}{P_1}\right]^{1-\frac{1}{n}}\right)} \quad \text{..............................} (4.9)$$

The analysis shows that, in the case of constant supply pressure (P_1), the nozzle's discharge increases as the pressure at its throat (P_2) falls. Yet, the discharge achieves a maximum when the nozzle pressure P_2 exceeds the critical value and after that, regardless of the pressure at the exit, the throat pressure and mass flow stay constant. By substituting the value of $\frac{P_2}{P_1}$ from equation (4.7) into equation (4.9), we can determine the maximum steam velocity at the nozzle's throat when the discharge is at its maximum.

$$V_{max} = \sqrt{2\left(\frac{n}{n-1}\right)P_1v_1\left(1 - \left[\left(\frac{2}{n+1}\right)^{\frac{n}{n-1}}\right]^{\frac{n-1}{n}}\right)}$$

$$V_{max} = \sqrt{2\left(\frac{n}{n-1}\right)P_1v_1\left(1 - \frac{2}{n+1}\right)}$$

$$V_{max} = \sqrt{2\left(\frac{n}{n-1}\right)P_1v_1\left(\frac{n-1}{n+1}\right)}$$

$$V_{max} = \sqrt{2\left(\frac{n}{n+1}\right)P_1v_1} \quad \text{..............................} (4.10)$$

- **Velocity of steam at the exit of nozzle in terms of heat drop can be written as**

$$V_2 = 44.72\sqrt{(h_1 - h_2)} \text{ m/s}$$

Here h_1, h_2 are the enthalpy (kJ/kg) at section 1 and section 2 respectively and V_2 is velocity (m/s) of steam at the exit.

4.7 Critical Pressure Ratio

A critical pressure can be defined as the pressure at which the area is minimum and discharge per unit area is maximum. The pressure ratio beyond which an increase in mass flow rate is not possible, even with a decrease in exit pressure, is known as the critical pressure ratio. This situation arises when the minimal cross section's Mach number is equal to one. So, the critical pressure ratio is the ratio which maximize the mass flow rate of discharged steam from a nozzle. This ratio can be expressed as

$$\frac{P_c}{P_i} = \left\{\frac{2}{n+1}\right\}^{\frac{n}{n-1}}$$

Here n is isentropic index and P_c, P_i are critical pressure and inlet pressure.

The critical pressure (P_c) is the pressure at the throat where the flow chokes, and the ratio of the critical pressure (P_c) at the choked plane (throat) to the inlet pressure (P_i) is known as the critical pressure.

For a perfect gas undergoing an adiabatic process the index 'n' is the ratio of specific heats k = c_p /c_v. There is no unique value for n as it changes with nature of steam or fluid. The critical pressure ratio for different steam conditions are following:

a) Critical pressure ratio for air (n = 1.4)

$$\frac{P_c}{P_i} = \left\{\frac{2}{n+1}\right\}^{\frac{n}{n-1}}$$

$$\frac{P_c}{P_i} = \left\{\frac{2}{1.4+1}\right\}^{\frac{1.4}{1.4-1}}$$

$$\frac{P_c}{P_i} = 0.528$$

$$P_c = 0.528P_i$$

b) Critical pressure ratio for saturated steam *(n = 1.135)*

$$\frac{P_c}{P_i} = \left\{\frac{2}{1.135+1}\right\}^{\frac{1.135}{1.135-1}}$$

$$\frac{P_c}{P_i} = 0.577$$

$$P_c = 0.577P_i$$

c) Critical pressure ratio for superheated steam *(n = 1.3)*

$$\frac{P_c}{P_i} = \left\{\frac{2}{1.3+1}\right\}^{\frac{1.3}{1.3-1}}$$

$$\frac{P_c}{P_i} = 0.546$$

$$P_c = 0.546P_i$$

d) Critical pressure ratio for wet steam *(n = 1.113)*

$$\frac{P_c}{P_i} = \left\{\frac{2}{1.113 + 1}\right\}^{\frac{1.113}{1.113-1}}$$

$$\frac{P_c}{P_i} = 0.582$$

$$P_c = 0.582P_i$$

4.7.1 Importance of critical pressure ratio

Suppose we have high pressure (P_1) steam in the two vessels X and Y connected via convergent nozzle as shown in figure. Let the pressure in the vessel Y is supposed to be P_2 which was earlier equal to P_1. Here in this situation there will be no flow. Due to this, the pressure will start gradually decreasing. The decrease in pressure initiates the discharge through nozzle which will increase slowly as shown in figure. At the point of time when the pressure P_2 reaches critical value the rate of discharge slowly reaches the maximum value. If pressure P_2 decreases and comes under critical value, then discharge rate will stay equal to value at critical pressure. This critical pressure will provide the high velocity (equal to velocity of sound i.e. sonic) to the steam. But in the convergent portion of the nozzle, the flow of steam is subsonic. In order to increase the steam velocity at supersonic level, the divergent portion is necessary. Here the steam is expanded below critical pressure to achieve desired velocity.

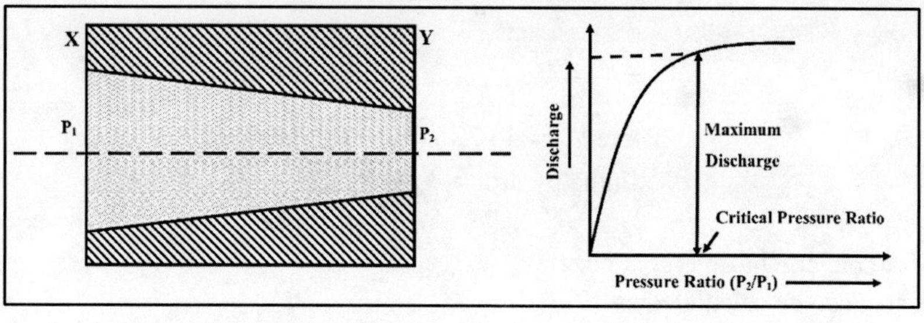

Fig. 4.3 Nozzle section and Discharge-Pressure curve

Thus to increase the velocity of steam above sonic velocity (supersonic), by expanding steam below critical pressure, divergent portion is necessary.

4.8 Methods of Calculation of Cross-Sectional Areas at Throat and Exit for Maximum Discharge

The cross sectional area at throat and exit of the nozzle can be calculated as:

a) For convergent nozzle

In case of convergent nozzle, the steam will exit from the throat. The area of the throat can be calculated by obtaining values of enthalpy drop, velocity at the throat, specific volume of steam at pressure P_2, and dryness fraction of the steam at pressure P_2.
Suppose P_1, P_2 are the entrance and exit pressure of the steam.
V_2 is the velocity at the throat while x_2 and V_{S2} are respectively the dryness fraction and specific volume of steam at pressure P_2.

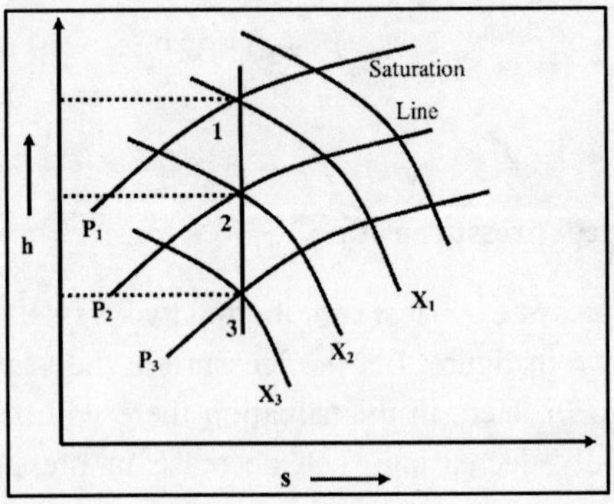

Fig. 4.4 Enthalpy (h) – Entropy (s) chart

Now mass flow rate can be calculated as:

$$m = \frac{A_2 \times V_2}{x_2 v_{s2}}$$

$$m = \frac{A_2 \times V_2}{x_2 v_{s2}}$$

$$A_2 = \frac{m.x_2 v_{s2}}{V_2} \quad \ldots\ldots\ldots\ldots\ldots\ldots\ldots (4.11)$$

Here $V_2 = 44.72\sqrt{(h_1 - h_2)}$ m/s

Enthalpy drop = $h_1 - h_2$ kJ/kg

b) For convergent-divergent nozzle

The exit area can be calculated as

$$m = \frac{A_3 \times V_3}{x_3 v_{s3}}$$

$$A_3 = \frac{m.x_3 v_{s3}}{V_3} \quad \ldots\ldots\ldots\ldots\ldots\ldots\ldots\ldots (4.12)$$

Here V_3 is the velocity at the throat while x_3 and v_{s3} are the dryness fraction and specific volume of steam at pressure P_3.

4.9 Effect of Friction in Nozzles

When the steam flows through the nozzle, it experiences friction with nozzle surface. Due to this friction, there is a small loss of steam's enthalpy, which is another term for total heat. As shown in Figure 4.5, this may be explained most efficiently with the use of an h-s diagram or a Mollier chart, which is described below:

- Initially, it is necessary to identify the point A that represents the initial conditions of the steam. It is a point where the saturation line and the initial pressure (P_1) line intersects.
- Now, in order to reach the line that represents the ultimate pressure (P_2), draw a vertical line across A. It occurs by assuming that the flow through the nozzle is isentropic, as indicated by the vertical line AB. The term "isentropic heat drop" refers to the term "heat drop" $(h_1 - h_2)$.

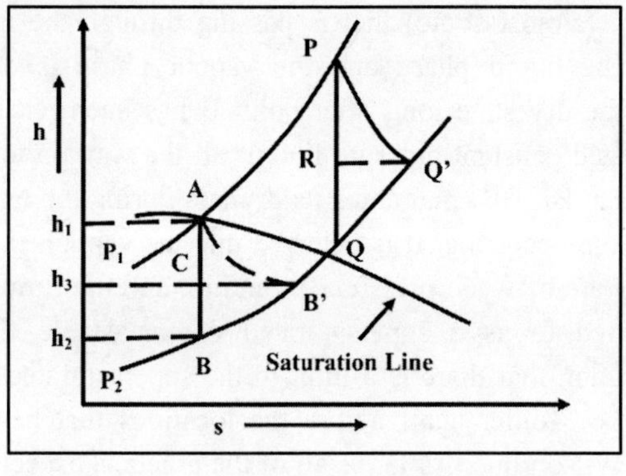

Fig. 4.5 Enthalpy (h) – Entropy (s) chart

- The actual heat loss in the steam will be smaller than (h_1 - h_2) because of the friction that occurs in the nozzle in the process. Let this heat decrease be shown as AC rather than AB.
- Since the expansion of steam comes to a halt at the pressure P_1, the ultimate condition of steam may be achieved by drawing a horizontal line through C and following it until it reaches the line that represents the final pressure (P_2) at B'.
- The curve AB', which represents adiabatic expansion, is now used to describe the real expansion of steam in the nozzle, rather of the classic curve AB, which represents isentropic expansion. The term "useful heat drop" refers to the actual heat decrease that occurred (h_1- h_3).

The ratio of the usable heat drop to the isentropic heat drop is known as coefficient of nozzle or nozzle efficiency, which is often symbolized by K.

Now, K = Useful heat drop/isentropic heat drop; K = AC/AB

$$K = \frac{h_1 - h_3}{h_1 - h_2} \quad \ldots\ldots\ldots\ldots\ldots\ldots\ldots (4.13)$$

4.10 Super Saturated Flow in Nozzles

As the steam expands in the nozzle, its pressure and temperature decrease, and it believed that the steam will begin to condense on contact with the saturation line. However, this is not a universal fact. Due to the high velocity, the duration that the steam remains in the nozzle is short, possibly not allowing sufficient time for the required heat transfer and formation of liquid droplets. As a result, the process of steam condensation is temporarily delayed. The phenomenon is referred to as super saturation, and the steam that occurs in the moist area without any liquid content is known as supersaturated steam.

Explanation on enthalpy-entropy chart:

It is easy to represent the isentropic expansion of superheated steam on the Mollier diagram with the help of line AE, as shown in Figure 4.6. The expansion occurs from supply pressure p_1 to back pressure p_b. A change in phase has to start at point B, which is the point at which the pressure line p_2 meets the saturation line, while expansion takes place. When it comes to nozzles, however, the process of condensation does not take place at point B under particular conditions. This is because the time which is available is very restricted due to the extremely high velocity of steam (almost sonic) that is passing through the nozzle. As a result, the equilibrium between the liquid phase and the vapour phase is delayed, and the vapor continues to grow in the dry state long after point B has been reached upon point B_1. By extending the superheated constant pressure line p_3 all the way up to B_1, one can figure out the pressure that exists at B_1. BB_1 maintains its dryness during the expansion process, which delays condensation from occurring. It is believed that the vapours will be supersaturated or supercooled when they are between pressures p_2 and p_3, and the flow of steam that occurs in nozzles will be referred to as a supersaturated or metastable flow of steam. Wilson established the observation that there is a limit to the supersaturated condition, and the line that he sketched on the Mollier chart across the locations that he observed subsequently became known as the Wilson line. As far as all of the practical aspects are involved, this line is now effectively the saturation line. As a result of the fact that the temperature of the vapour is always lower than the saturation temperature that corresponds to that pressure, the flow is also referred to as supercooled flow. This is the case regardless of the pressure that falls between p_2 and p_3. The degree of undercooling refers to the difference in temperature between these two temperatures. Condensation takes place at constant enthalpy and pressure, as seen by the horizontal line BC, when the expansion reaches point B on the Wilson line. Furthermore, the pressure remains constant throughout the condensation process. CD is a representation of the further isentropic expansion that occurs to the exit pressure point. One definition of the term "degree of supersaturation" describes it as the ratio of the saturation pressures that correspond to the temperatures at B and B_1.

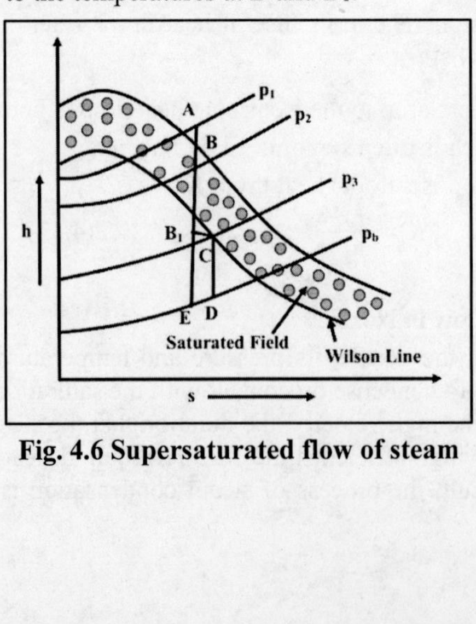

Fig. 4.6 Supersaturated flow of steam

4.11 Effects of Supersaturation

The phenomenon of super saturation takes place in the nozzle as a result of a delay in condensation. Because of this, the mass flow rate that initiates from the nozzle is greater than the value that was scheduled for it. The following are the key effects in nozzle which supposed to undergo super saturation:

- The super saturation decreases the rate at which heat is lost under the same pressure conditions.
- Supersaturated expansion does not result in condensation, hence the temperature at which super saturation occurs will be lower than the saturation temperature corresponding to the pressure. Consequently, the density of supersaturated steam will be greater than that of steam in equilibrium, resulting in an increase in the mass of released steam.
- It leads to an increase in both the entropy and specific volume of the steam.
- The super saturation also increases the dryness fraction.

4.12 Steam Injector

A steam injector is a specialized instrument used to inject water vapour into an alternate medium, frequently liquid water. Steam injectors have little application in thermal engineering since they need to be placed close to the steam generator to function properly. Primarily, these injectors were used as a component in steam boilers or in systems that harness the thermal energy of surplus water vapour via heated water.

4.12.1 Working steam jet injector

A steam injector can be explained as an application of the same principle that supports a steam nozzle. It works by utilizing the kinetic energy of a steam jet, which in turn increases the pressure and velocity of the water. One of its primary functions is to provide pressure to the feed water in order to force it into steam boilers. The Figure 4.7 gives an illustration of the operation of a steam injector. With the use of a convergent nozzle at 1, the steam which is produced by the boiler is increased to a high velocity and expands while passing through it. The steam jet makes its way into the mixing cone, where it transfers its velocity to the water supply that is coming from the feed tank. The steam gets condensed because of the cold water. The jet created at 2 as a consequence of the steam and water is at the atmospheric pressure with a significant amount of velocity. In the subsequent step, the mixture is introduced into the delivery pipe at 3 by means of a diverging cone or diffuser. During this process, the kinetic energy is decreased and transformed into pressure energy. The pressure energy that is available is sufficient to overcome the pressure that is present in the boiler and to elevate the water through a height H_b. This means that the pressure of the water as it exits the delivery pipe has to be about 20% greater than the pressure of the boiler in order to overcome all resistances. During the process of initiating the injector, there is an exit that is located in the space between the mixing cone and the diverging cone. This allows any extra water to seep out of the gap.

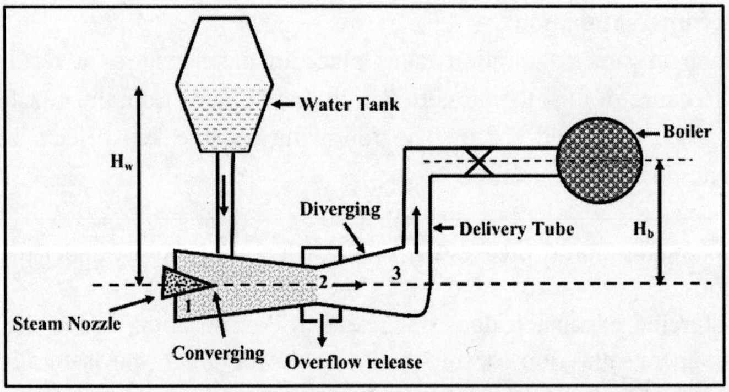

Fig. 4.7 Schematic diagram of Steam Injector

- **Mass of injected water**

As per principle of conservation of momentum, the sum of momentum of steam at the entry and momentum of water at the entry will be equal to the mixture of steam and water at the exit. Let's suppose the steam mass is of 1kg. Now,

$$1 \times V_s + M_w V_w = \{1 + M_w\} V_m$$

$$V_s - V_m = M_w \{V_m + V_w\}$$

$$M_w = \frac{V_s - V_m}{V_m + V_w} \quad \text{........................ (4.14)}$$

Here M_w is the mass of water entering the nozzle in kg/kg of steam and V_w, V_s are velocity of water entering the nozzle in m/s and velocity of steam leaving the nozzle in m/s respectively. V_m is the velocity of mixture leaving nozzle at 2 in m/s.

- **Steam velocity at point 1**

The steam velocity at point 1 can be calculated by assuming that expansion of steam is isentropic. It can be expressed as

$$V_s = 44.72\sqrt{(h_1 - h_2)}$$

Here h_1-h_2 is the enthalpy drop.

- **Steam velocity at point 2**

The total energy per kg of water at point 2 will be = $\frac{P_m}{w} + \frac{V_m^2}{2000}$

Here P_m is the pressure of the mixture leaving the nozzle at 2 in kPa and w is density of the mixture at 2 in kg/m^3.

This energy will be used to lift the water at height H_b meter to transfer it to the boiler. The final pressure should be greater than sum of H_b and boiler pressure. Suppose V is the velocity of fluid in delivery tube. Then,

The total energy per kg of water will be = $\frac{P}{w}+\frac{V^2}{2000}+\frac{gh_b}{1000}$

Here P is the absolute pressure of the steam in the boiler in kPa. Then,

$$\frac{P_m}{w}+\frac{V_m^2}{2000}=\frac{P}{w}+\frac{gh_b}{1000}+\frac{V^2}{2000}$$

$$V_m=\sqrt{2000\left\{\left(\frac{P-P_m}{w}\right)+\frac{gh_b}{1000}+\frac{V^2}{2000}\right\}}$$

When the water reaches the boiler, the kinetic energy ($\frac{V^2}{2000}$) becomes equal to the pressure energy by addition of 9.3 m (approx.) as lift h_b and velocity becomes 13.5m/s. Therefore,

$$V_m=\sqrt{2000\left\{\left(\frac{P-P_m}{w}\right)+\frac{gh_b}{1000}+\frac{g}{1000}\times 9.3\right\}} \quad \text{..................... (4.15)}$$

- **Area of nozzles**

Let A_1, A_2 be the area of steam nozzle at 1 and 2; V_s, V_1 be the velocity of steam leaving the converging cone and specific volume of steam after expansion in nozzle at 1 and m, m_w be the mass of water required to be delivered in kg/s, and mass of water entering the mixing cone in kg/kg of steam.

Now, mass of supplied steam, $m_s=\frac{A_1V_s}{V_1}=\frac{m}{m_w}$

Or $$A_1=\frac{m_sV_1}{V_s}$$

$$A_1=\frac{V_1}{V_s}\times\frac{m}{m_w}$$

$$A_2=\frac{m_s+m}{1000V_m} \qquad (\because m_s+m=A_2V_m\rho)$$

Here ρ is the density which is equal to 1000 kg/m^3.

<u>*Solved Examples*</u>

Numerical 4.1 A stream of dry saturated steam, initially at a pressure of 15 bar, is injected into a nozzle and then discharged at a pressure of 1.5 bar. Determine the final velocity of the steam when the initial velocity of the steam is negligible. Determine the percentage of the final velocity which decreases if friction caused 10% of the heat drop to be lost.

Solution: Given: p_1 = 15 bar and p_2= 1.5 bar

The enthalpy of dry saturated steam from the steam table,

$$h_1 = 2789.9 \text{ kJ/kg}$$

$$h_2 = 2693.4 \text{ kJ/kg}$$

The enthalpy drop will be: h_1-h_2= 96.5 kJ/kg

Final velocity of the steam can be given as $V_s = 44.72\sqrt{(h_1 - h_2)}$

$$V_s = 44.72\sqrt{(96.5)}$$

$$V_s = 439.3\ m/s$$

heat drop to be lost = 10% or 0.1

Nozzle efficiency (K) will be = 1-0.1 = 0.9

Final velocity of the steam can also be given as $V_s = 44.72\sqrt{\mathrm{K}(h_1 - h_2)}$

$$V_s = 44.72\sqrt{(0.9 \times 96.5)}$$

$$V_s = 416.8 \text{ m/s}$$

Therefore, final reduction will be = $\frac{439.3-416.8}{439.3} = 0.051\ or\ 5.1\%$

Numerical 4.2 The dry saturated steam, initially at a pressure of 10 bar, undergoes expansion in a nozzle to reach a final pressure of 0.7 bar. If friction is ignored, determine the nozzle's steam output velocity and dryness fraction using the Mollier diagram.

Solution: Given: p_1 = 10 bar and p_2= 0.7 bar

By using Mollier diagram, we can obtain dryness fraction. The procedure of using Mollier diagram is as follow:

- Start with locating point A on the saturation line where initial pressure meets.
- Due to isentropic expansion, meet the final pressure 0.7 bar at point by drawing vertical line through A.

Now, from the Mollier diagram, $h_1 = 2772\ kJ/kg$ and $h_2 = 2310\ kJ/kg$

Now, the heat drop will be = $h_1 - h_2$= 462 kJ/kg

Final velocity of the steam can be given as $V_s = 44.72\sqrt{(h_1 - h_2)}$

$V_s = 44.72\sqrt{(462)}$ = 961 m/s

The dryness fraction can be obtained through Mollier diagram and found as 0.848.

Numerical 4.3 The convergent diverging nozzle is fed with steam at a pressure of 10 bar and a temperature of 210°C. The throat area of the nozzle is 15 cm^2. The exit is under conditions of critical pressure. Determine the discharge coefficient for a flow rate of 7200 kg of steam per hour.

Solution: Using steam table, the specific volume of steam (in case of superheated steam) is = 0.2113 m³/kg.

The value of n of superheated steam will be 1.3 (∵ steam becomes superheated when the temperature is above 100°C at normal atmospheric pressure)

$$m_{max} = A \times \sqrt{1000\, n \times P_1/v_1 \left(\frac{2}{n+1}\right)^{\frac{n+1}{n-1}}}$$

$$m_{max} = 15 \times 10^{-4} \sqrt{1000 \times 1.3 \times \frac{(10 \times 10^2)}{0.2113}\left(\frac{2}{1.3+1}\right)^{\frac{1.3+1}{1.3-1}}}$$

$$m_{max} = 2.177$$

Now, discharge coefficient = Actual/Theoretical = $\frac{7200}{2.177 \times 3600} = 0.92$

Numerical 4.4 The convergent diverging nozzle receives air at a pressure of 20 bar and a temperature of 18°C. The nozzle has a throat diameter of 1.25 cm and releases the air into the atmosphere. The adiabatic index of air is 1.4, while the characteristic constant is 287. Determine the mass of air released per minute.

Solution: η_{air} = 1.4 and C_p for air = 1.005 J/kg-K

We know $\frac{P_2}{P_1} = \left(\frac{2}{n+1}\right)^{\frac{n}{n-1}}$ and for $\frac{T_2}{T_1} = \left(\frac{2}{n+1}\right)^{\frac{n}{n-1} \times \frac{n-1}{n}}$

Now $\frac{T_2}{T_1} = \left(\frac{2}{n+1}\right)$ and $T_2 = \frac{2}{1.4+1} \times (273+18)$

$$T_2 = 242.5\ K$$

$$\frac{P_2}{P_1} = 0.528\ (for\ air)$$

$$P_2 = P_1 \times 0.528 = 10.56\ bar\ or\ 10.56 \times 10^5\ pascal$$

The specific volume at the throat can be calculated as $v_2 = \frac{T_1 \times Charateristic\ Constant}{P_2}$

$$v_2 = \frac{242.5 \times 287}{10.56 \times 10^5} = 0.0659\ \text{m}^3/\text{kg}$$

Now, Velocity at throat will be, $V_2 = 44.72\sqrt{C_p(T_1 - T_2)}$

$V_2 = 44.72\sqrt{1.005(291 - 242.5)}$ = 312.12 m/s

Mass of air discharged per minute is $m = \frac{A_2 V_2}{v_2} \times 60$

$$m = \frac{A_2 V_2}{v_2} \times 60$$

$$m = \frac{0.784 \times (0.0125^2) \times 312.127}{0.0659} \times 60$$

$$m = 34.81\ kg/min$$

Numerical 4.5 The initial pressure of the dry saturated steam is 11 bar as it undergoes expansion in a set of convergent divergent nozzles at a rate of 5 kg/s. There is a 1.5 bar back pressure. The nozzle efficiency is 90% and the approximate cross-sectional area of each nozzle's throat is about 1 cm^2. Determine the ideal number of nozzles and precise dimensions for the throat and exit areas. The index for expansion is 1.119.

Solution: We know $\frac{P_2}{P_1} = \left(\frac{2}{n+1}\right)^{\frac{n}{n-1}}$

$$P_2 = P_1\left(\frac{2}{n+1}\right)^{\frac{n}{n-1}}$$

$$P_2 = 11\left(\frac{2}{1.119+1}\right)^{\frac{1.119}{1.119-1}}$$

$$P_2 = 6.4\ bar$$

Using enthalpy-entropy chart, h_1= 2781.7 kJ/kg (value at 11 bar)

h_2= 2675 kJ/kg (value at 6.4 bar); v_{s2}=0.2970 m^3/kg; $x_2 = 0.96$

h_3= 2445 kJ/kg (value at 6.4 bar); v_{s3}=1.1593 m^3/kg; $x_2 = 0.885$

$$V_2 = 44.72\sqrt{(h_1 - h_2) \times \eta_n}$$

$$V_2 = 44.72\sqrt{(2781.7 - 2675) \times 0.9}$$

$$V_2 = 438.23\ m/s$$

$$m = \frac{A_2V_2}{v_2} = \frac{1\times10^{-4}\times438.23}{0.2970\times0.96} = 0.153\ \text{kg/s}$$

Now, the number of nozzles are $\frac{5}{0.153} = 33\ (approx.)$

$$A_3 = \frac{mv_3}{V_3} = \frac{mx_3v_{s3}}{V_3}$$

$$A_3 = \frac{5\times0.885\times1.1593\times10^{-4}}{778.4\times33} = 1.996\ \text{cm}^2$$

Numerical 4.6 The steam, initially at a pressure of 10.5 bar and with a dryness fraction of 0.95, undergoes expansion through a convergent-divergent nozzle, leading to a final steam pressure of 0.85 bar. For the conditions of maximum discharge, find the throat's velocity. The index for expansion might be considered as 1.135. Determine the mass flow rate of steam passing through the nozzle.

Solution: For maximum discharge, the pressure will be $\frac{P_2}{P_1} = \left(\frac{2}{n+1}\right)^{\frac{n}{n-1}}$

$$P_2 = 10.5\left(\frac{2}{1.135+1}\right)^{\frac{1.135}{1.135-1}}$$

$$P_2 = 6.06\ bar$$

Now the velocity at the throat can be written as

$$V_2 = \sqrt{2(^{n}/_{n+1})P_1 v_1}$$

$$V_2 = \sqrt{2(\frac{1.135}{1.135+1}) \times 10.5 \times 10^5 (0.95 \times 0.185)}$$

$$V_2 = 443\ m/s$$

Also, $P_1 v_1^n = P_1 v_2^n$

$$10.5(0.95 \times 0.185)^{1.135} = 6.06 \times (v_2^{1.135})$$

$$v_2 = 0.285\ m^3/kg$$

The mass flow rate can be obtained as $\dot{m} = \frac{A_2 V_2}{v_2}$

$$\dot{m} = \frac{1 \times 443}{0.285}$$

$$\dot{m} = 1554.3 kg/m^2$$

Numerical 4.7 Determine the critical pressure and throat area per unit mass-flow rate of steam as it undergoes expansion through a convergent-divergent nozzle, starting from an initial pressure of 10 bar and reaching atmospheric pressure of 1 bar. Let's assume that the initial velocity at inlet is very small and that the expansion process is isentropic.

Solution: The critical pressure can be calculated as

$$\frac{P_c}{P_i} = \left\{\frac{2}{n+1}\right\}^{\frac{n}{n-1}}$$

$$\frac{P_c}{P_i} = \left\{\frac{2}{1.135+1}\right\}^{\frac{1.135}{1.135-1}}$$

$$\frac{P_c}{P_i} = 0.577$$

$$P_c = 0.577 \times 10$$

$$P_c = 5.77\ bar$$

The throat area will be $A_2 = \frac{m.v_2}{V_2}$ $or\ A_2 = \frac{m.x_2 v_{s2}}{V_2}$

Here V2 is the velocity at the throat while x_2 and Vs2 are the dryness fraction and specific volume of steam at pressure P2.

Using Mollier chart and steam table, at pressure 5.77 bar, h_1= 2778, h_2 = 2675kJ/kg, x_2= 0.962 and v_{s2}= 0.328 m^3/kg
Now, specific volume at throat will be $v_2 = x_2 v_{s2}$
$v_2 = 0.962 \times 0.328 = 0.316\ m^3/$kg

The exit velocity of steam can be calculated as $V_2 = 44.72\sqrt{(2778 - 2675)}$ m/s

$V_2 = 453.87$ m/s

The area can be calculated as $A_2 = \frac{m.v_2}{V_2}$

$$A_2 = \frac{1 \times 0.316}{453.87}$$
$$A_2 = 696\ mm^2$$

UNIT SUMMARY

Based on the contents covered in this unit you should remember that:

- A critical pressure can be defined as the pressure at which the area is minimum and discharge per unit area is maximum.
- The critical velocity is the steam velocity at the throat that corresponds to the maximum discharge condition.
- In a wet region, supersaturation refers to the presence of fluid in a superheated state.
- A steam nozzle may be defined as a passage of varying cross-section through which the heat energy of steam is converted to kinetic energy.
- The nozzle efficiency is defined as the ratio of the actual enthalpy drop to the isentropic enthalpy drop between the same pressures.
- A steam injector is a device employed to force water into the boiler under pressure.
- The nozzle can achieve the maximum discharge value by calculating the ratio between the steam pressure at the throat and the initial pressure.
- The coefficient of nozzle or nozzle efficiency, often represented by K, is the ratio of the usable heat drop to the isentropic heat drop.
- The phenomenon of super saturation occurs in the nozzle due to a condensation delay.

Exercises

Multiple Choice Questions

1. What type of flow is typically assumed in the analysis of steam flow through nozzles?

   A) Isothermal flow

   B) Isentropic flow

   C) Isobaric flow

   D) Adiabatic flow

2. Which of the following equations is used to describe the relationship between steam properties in a nozzle?

   A) Bernoulli's equation

   B) Continuity equation

   C) Energy equation

   D) Both A and B

3. The efficiency of a steam nozzle can be affected by which of the following?

   A) The angle of the nozzle exit

   B) The temperature of the steam

C) The pressure drop across the nozzle

D) All of the above

4. In the context of the Mollier chart, which property is used to determine the velocity of steam at the nozzle exit?

A) Enthalpy

B) Pressure

C) Temperature

D) Specific volume

5. When applying the Mollier chart to find the exit velocity of steam from a nozzle, what is the first step?

A) Determine the steam quality

B) Find the enthalpy at the nozzle inlet and outlet

C) Calculate the temperature drop

D) Measure the pressure drop across the nozzle

6. In a converging-diverging nozzle, the steam flow reaches sonic speed at:

A) The converging section

B) The diverging section

C) The nozzle throat

D) The exit of the nozzle

7. If the inlet steam conditions are 8 bar and 400°C and the exit pressure is 1 bar, what happens to the velocity of the steam as it exits the nozzle?

A) The velocity decreases

B) The velocity increases

C) The velocity remains constant

D) The velocity first increases and then decreases

8. Friction in a nozzle primarily affects which parameter of the steam flow?

A) Mass flow rate

B) Enthalpy

C) Exit velocity

D) Pressure drop

9. In the presence of friction, the actual enthalpy drop across a nozzle is generally:

A) Greater than the theoretical enthalpy drop

B) Less than the theoretical enthalpy drop

C) Equal to the theoretical enthalpy drop

D) Independent of the theoretical enthalpy drop

10. Which of the following parameters most directly affects the performance of a steam jet injector?

    A) The steam pressure

    B) The steam temperature

    C) The length of the nozzle

    D) The color of the injector

11. What is the primary purpose of a steam jet injector?

    A) To heat the steam before it enters the turbine

    B) To pump fluids using the energy of steam

    C) To measure steam pressure

    D) To cool down the steam

12. In a steam jet injector, the steam's kinetic energy is converted into:

    A) Thermal energy

    B) Chemical energy

    C) Mechanical energy

    D) Pressure energy

Answer Key:

1 B	2 D	3 D	4 A	5 B	6 C
7 B	8 C	9 B	10 A	11 B	12 D

Short and Long Answer Type Questions

- Why does the nozzle area usually taper or converge?
- What does the term "isentropic expansion" mean when addressing the flow of steam through a nozzle?
- How is the heat drop technique used to compute the steam velocity at the nozzle's exit?
- How does the steam velocity in a nozzle relate to the enthalpy drop?
- What data is required in the Mollier chart in order to calculate the steam exit velocity from a nozzle?
- How can you determine the exit steam velocity if the Mollier chart shows the inlet enthalpy as 2900 kJ/kg and the exit enthalpy as 2500 kJ/kg?
- How does the isentropic relation help in calculating the cross-sectional area at the nozzle exit?
- Define the steam turbine nozzle's critical pressure ratio. Determine its value analytically with respect to the expansion index.
- What is the effect of friction on the flow rate through a steam nozzle? Provide an explanation using the h-s diagram.
- What do you mean by a supersaturated flow? Provide an explanation using the h-s diagram.

Know More

- ✓ The expansion of steam in steam turbine is isentropic i.e. it takes place at constant entropy. This hypothesis is necessary in numerous theoretical evaluation of nozzle performance, leading to the interpretation that the process is both reversible and adiabatic.
- ✓ Sometimes the flow in the nozzle is referred as choked as the velocity of steam exceeds the speed of sound. At this point, the mass flow rate has no effect of pressure downstream. This choked flow is key concept which plays crucial role in the designing of steam nozzles.
- ✓ The efficiency of steam turbine can be determined by obtaining ratio of actual decrease in enthalpy to the ideal decrease in enthalpy during isentropic expansion.
- ✓ In some industrial applications, there may be a partial recovery of steam pressure as the flow leaves the nozzle. This phenomenon is caused by complex interactions between the flow and the shape of the nozzle.
- ✓ The temperature of the steam has an effect on its density and viscosity, which successively impacts the design of the nozzle. The presence of high-temperature steam could cause material challenges as a result of thermal expansion and tension, thereby requiring the use of advanced materials.
- ✓ Latest nozzle design usually depends upon computational fluid dynamics (CFD) simulations to accurately predict performance. These simulations include complex flow phenomena, such as turbulence and non-ideal effects, providing valuable insights that are hard to get simply through analytical methods.

References and Suggested Readings

- ➢ Thermal Engineering by B. R. Sarkar (McGraw-Hill Education (India) Pvt Limited).
- ➢ Thermal Engineering by R.K. Rajput (Laxmi Publications (P) Ltd).
- ➢ A Textbook of Thermal Engineering by R.S. Khurmi & J.K. Gupta (S. Chand & Company Ltd).
- ➢ Thermal Engineering by Mahesh. M. Rathore (McGraw-Hill Education (India) Pvt Limited).

Dynamic QR Code for Further Reading

1. NPTEL Lectures on "Use of steam nozzles and flow analysis" by Dr P.K. Mondal, Department of Mechanical Engineering, Indian Institute of Technology Guwahati, India

2. Lectures on “Steam nozzles using mollier diagram” by Basic Engineering Concepts YouTube channel.

3. NPTEL Lectures on “Converging & Diverging Nozzle Operation” by Prof. Srisha Rao M.V, Department of Aerospace Engineering, Indian Institute of Science, Banglore, India.

5 Steam Turbines

UNIT SPECIFICS

This unit presents information related to the following topics:

- Classification of steam turbines and comparison between impulse & reaction turbines.
- Working and velocity diagrams of a simple De-lavel turbine.
- Methods of reducing rotor speed and compounding for velocity and pressure.
- Working and velocity diagrams of a Parson's Reaction turbine
- Bleeding, re-heating, re-heating factors & governing of steam turbines.

This unit gives readers in depth knowledge of steam turbines. The unit includes classification of the turbines and from the classification, the comparison between impulse and reaction turbine is discussed. The mechanism of steam turbines can be learned with the help of velocity diagrams in which flow of steam is analysed. The required formulae for work done, axial thrust, tangential thrust, blade and diagram efficiency, stage efficiency, nozzle efficiency are explained effectively. The various methods of reducing rotor speed, compounding for velocity are discussed in detail to provide comprehensive knowledge to the readers. This unit also includes comprehensive information on governing of steam turbines, aiming to give readers a concise and efficient understanding of the topic.

The unit completes with a brief review of the key concepts. Furthermore, the unit includes a significant portion of multiple-choice and descriptive questions. The unit includes a list of readings and sources for students to utilize for practice. It is important to mention that QR codes have been included in several parts to give more information on various areas of interest. The readers can scan these codes to obtain pertinent and corroborating information. The unit emphasizes the inclusion of video resources and QR codes to provide more information on numerous areas of interest. These resources can be accessed and viewed on mobile phones by browsing or scanning QR codes.

RATIONALE

The steam turbine has emerged as an essential device in the generation of energy. It is a device that utilizes the thermal energy from high-pressure steam and converts it into mechanical work by means of a rotating output shaft. This unit will help the readers to know the important aspects of steam turbine. The classification of steam turbine helps in understanding the different types of turbines and their working mechanism. The governing is an important part of turbine functioning as it maintains constant speed under all conditions. This unit will also help the readers to understand the fundamental of governing of steam turbines.

PRE-REQUISITES

Before reading this unit, the students are advised to revisit the following:
Basic Mechanical Engineering (MEPC102)

UNIT OUTCOMES

After studying this unit students will be able to:

U5-O1: Classify the steam turbine and effectively differentiate among impulse and reaction turbines.

U5-O2: Learn about the simple De-lavel turbine with velocity diagrams.

U5-O3: Understand the various methods of reducing rotor speed and compounding for velocity.

U5-O4: Learn about the Parson's reaction turbine and its velocity diagrams.

U5-O5: Understand the governing of steam turbines and related numerical.

Unit 5 Outcomes	EXPECTED MAPPING WITH COURSE OUTCOMES (1- Weak Correlation; 2- Medium Correlation; 3- Strong Correlation)				
	CO-1	**CO-2**	**CO-3**	**CO-4**	**CO-5**
U5-O1	2	1	-	-	3
U5-O2	2	-	-	1	3
U5-O3	1	-	-	1	3
U5-O4	2	1	-	1	3
U5-O5	1	2	-	1	3

5.0 Steam Turbines

The history of the steam turbines dated back to the first century. The hero of Alexandria writes about the ancient Aeolipile, a toy that dated to around the first century in Roman Egypt, and which is possibly the first example of a reaction steam turbine. Later on, in 1884, Charles Parsons created the first steam turbine, which produced 7.5 kilowatts (10.1 horsepower) of energy when it was coupled to a dynamo. The Parson's steam turbine which was a reaction turbine, changed naval combat and sea transportation while providing cheap, abundant power. In collaboration with John Brown & Company, the International Curtis Marine Turbine Company in the United States first designed and patented the impulse-type Brown-Curtis turbine in the 1900s. It was designed for warships and commerce vessels (particularly for cruise ships and Royal Navy combatants) using John Brown engines.

In simple words, "A steam turbine is a device which converts thermal energy to the mechanical work for rotating electrical generator to generate electricity".
The steam turbine operates through following steps:

- The steam turbine uses a heat source for heating the water and then produces steam. The heat source can be of many types such as coal, sunlight or nuclear fuel.
- The above-mentioned procedure enables the water molecule to expand.
- The generated steam then moves towards the blades of the turbine to convert thermal energy into kinetic energy due to rotation. The turbines can decrease the pressure of steam by keeping small but continuous increments. This approach helps in improving the efficiency of turbine and improves the electricity generation.
- The electricity is generated through a generator which is coupled to the steam turbine.

5.1 Classification of Steam Turbines with Examples

Steam turbines use the kinetic energy of steam to produce rotational motion. They are primarily classified according to the method of action, type of flow, and the number of stages. The further classification of the turbines is shown in the figure 5.1.

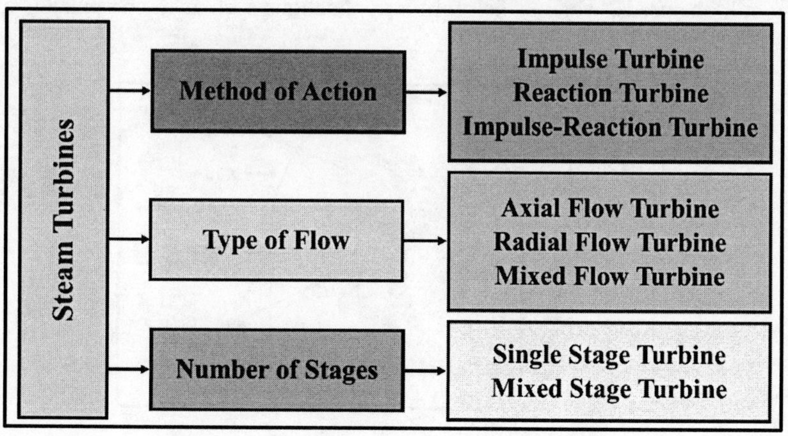

Fig. 5.1 Classification of steam turbines

An **impulse turbine** is a turbine which transforms kinetic energy of a working fluid in to mechanical energy. The obtained mechanical energy is then used to run a generator to produce electricity. A **reaction turbine** is a turbine which transforms potential energy of a working fluid in to mechanical energy which is further used to produce electricity through generator. This turbine is mostly used in hydroelectric power plants. The key difference among these two is that the impulse turbine relies on strike of high kinetic energy of the working fluid on blades whereas reaction turbine uses both potential as well as kinetic energy of the working fluid. The reaction turbines have better efficiency and can work in variable fluid flow. Steam expands continuously in both fixed and moving blades in an **impulse-reaction turbine** as it passes over them. The pressure loss is a continuous phenomenon that affects both fixed and moving blades.

The **axial flow turbine** is a turbine in which the direction of the fluid or water flow through the blades aligns with the turbine shaft's rotation whereas in **radial flow turbine**, the flow of the fluid or water is radial to the shaft. In a **mixed flow turbine**, the fluid enters in a radial direction and exits in axial direction.

The steam expands only once as it moves through the turbine in a **single-stage steam turbine**. This indicates that all of the steam's energy is extracted in a single step. But in **multistage steam turbine**, the steam travels through many stages of rotating and stationary blades each of which extracts some energy from the steam.

5.1.1 Principle of Impulse turbine

An impulse turbine (Figure 5.2) is a type of steam turbine that generates mechanical energy by the impact of high-velocity steam jets. It is based on the Newton's third law of motion, which states that every action has an equal and opposite reaction. Impulse turbines have some benefits such as it is easy to operate in smaller power production systems because of simple design and operation. Their ability to transfer up to 70% of thermal energy into mechanical energy is another reason for their great efficiency. Also, impulse turbines are capable of operating at high speeds, which is helpful for applications such as the propulsion systems of airplanes.

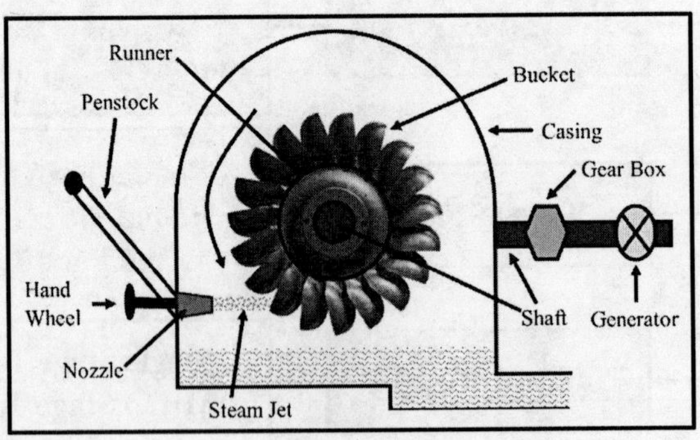

Fig. 5.2 Impulse Turbine

The conversion of the kinetic energy of a high-velocity jet of steam into mechanical energy is the fundamental concept that describes the operation of an impulse turbine. One of the components of the turbine is a fixed set of nozzles that are responsible for directing the steam onto a set of blades that are installed on a rotor. It is the purpose of the blades of an impulse turbine to reroute the flow of steam, which in turn causes the rotor to perform rotation. The steam is accelerated to a very high velocity as it travels through the nozzle, which results in the formation of a jet of steam that travels at a very high speed. The force of the steam then causes the rotor to rotate as a result of the impact of the steam jet on the blades of the rotor. After that, the steam is expelled from the turbine by means of the exhaust system.

The blades of an impulse turbine are constructed in such a way that they make sure the steam jet strikes them at the most suitable angle. This results in the rotor being subjected to the greatest possible amount of force. It is possible to do this by designing the blades in a way that enables the steam to shift direction without experiencing a significant loss in velocity. The quantity of power that is produced by an impulse turbine is dependent on a number of specific parameters. These parameters include the mass flow rate of the steam, the velocity of the steam, and the number of stages that are included inside the turbine. It is possible to collect additional energy from the steam by using multi-stage impulse turbines. These turbines work by directing the steam through a sequence of nozzles and blades, with each step extracting a small amount of the available total energy.

Components of Impulse Turbine

The operation of an impulse turbine is dependent on a number of different components, each of which plays a significant part in the overall operation of the turbine. In an impulse turbine, the following are the primary components:

- **Nozzle:** The first component of an impulse turbine is known as the nozzle. The movement of the steam onto the blades of the turbine is under the control of these fixed components. The function of the nozzles is to increase the velocity of the steam in order to transform the potential energy of the steam into kinetic energy.
- **Blades:** The blades are connected to the rotor and extract energy from the high-velocity steam. It is the function of the blades to change the direction of the steam flow in order to transform the kinetic energy of the steam into rotational energy. According to the design of the turbine, the blades may have a variety of shapes, such as radial, axial, or tangential rotations.
- **Rotor:** A shaft is attached to the rotor, which is the component of the impulse turbine that rotates. There are a number of blades that constitute the rotor, and are responsible for extracting energy from the high-velocity steam. They convert high velocity steam into rotational energy. After that, the energy that is generated by the rotation is transmitted to the shaft, which is then linked to a generator.
- **Casing:** The casing houses the rotor, blades, and nozzles of the impulse turbine. It is the outermost shell of the turbine. The purpose of this device is to hold the steam and to send it through the nozzles and onto the blades. In most cases, the casing is constructed out of steel or cast iron, and it is intended to be able to endure the high pressures and temperatures that are often associated with steam.
- **Bearing:** Bearings sustain the rotor and enable it to spin freely. Most of the time, they are made out of metal and are intended to resist tremendous weights and speeds.
- **Governor:** The governor is a control mechanism that is responsible for regulating the speed at which the turbine operates. It does this by regulating the quantity of steam that is pushed through the nozzles, which allows it to keep the speed constant.

5.1.2 Principle of reaction turbine

A reaction turbine (Figure 5.3) is a kind of turbine that generates torque by reacting to the pressure of a liquid, mass, or gas. By using Newton's third rule of motion, which stipulates that a given action results in an equal reaction in the opposite direction, it is possible to execute the operation of a reaction turbine. The flow of water over the fixed blades of the turbine leads to the generation of force by the turbine.

In a reaction turbine, the turbine's rotating blades are intended to enable the steam to expand and decrease in pressure as it flows over them. Due to the decrease in pressure on the moving blade, a reactive force will be generated. This force will cause the blades to rotate. Unlike the impulse turbine, the reaction turbine does not use nozzles to control the flow of steam. As an alternative, the blades that extend radially from the rotor's outer edge are placed and formed such that the cross-sectional shape between the blades appears like a nozzle. The blades are attached to the rotating section of the rotor and are referred to as the moving blades. The fixed blades, which have a similar shape as the rotating blades, are attached to the outer casing of the rotor and installed to direct the flow of steam towards the rotating blades. A reaction turbine is propelled by three primary forces. The reactive force is generated on the rotating blades as they expand and accelerate while passing through the nozzle-shaped gaps between them. Secondly, the reactive force is generated on the rotating blades when the steam flows through and changes direction. Furthermore, the rotation of the reaction turbine is more enhanced by the impact force of the steam on the blades.

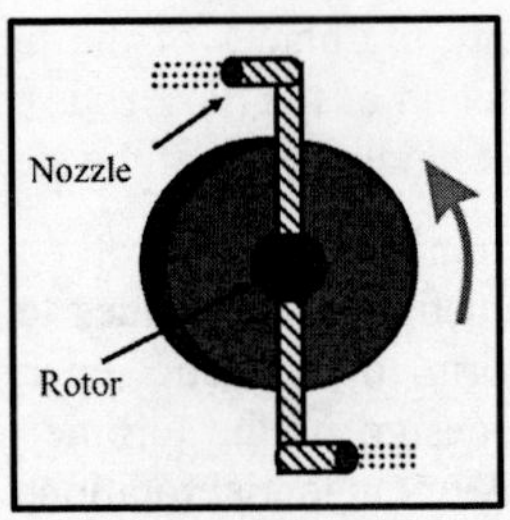

Fig 5.3 Reaction Turbine

The main parts of Reaction Turbine include:

- **Spiral Casing**

The spiral casing is responsible for evenly dispersing steam from a pipeline across the guiding ring. The cross-sectional area of the casing decreases progressively from the entry to the tip, creating a spiral or scroll shape. This design facilitates the optimal movement of steam.

- **Guide Vanes**

The guiding vanes are positioned inside the spiral casing, mounted in a wheel-like configuration between two rings. The purpose of these vanes is to facilitate the entry of water into the turbine while minimizing the occurrence of shocks or eddies. The regulating shaft, driven by a governor, allows for precise control over the amount of steam flowing into the turbine.

- **Runner and Runner Blades**

The turbine runner of a turbine may have blades that are either attached to a shaft or a ring, depending on the kind of turbine. These blades are meticulously engineered to provide seamless entry and exit without inducing abrupt disturbances. The choice of material for the runner, whether it be cast iron or steel alloys, depends on the specific working circumstances.

- **Draft Tube**

A draft tube is a component used to improve the efficiency of the turbine by reducing the velocity of the steam leaving the turbine. Upon traversing the turbine runner, the steam descends into the draft tube. The draft tube amplifies the steam head by a magnitude equivalent to the elevation of the runner outlet above the tailrace, hence increasing the efficiency of the turbine.

5.1.3 Principle of Impulse-Reaction turbine

The impulse-reaction turbine operates based on both the impulse principle and the reaction principle. In this type, both moving blades and stationary nozzles experience a pressure decrease. The steam experiences an increase in kinetic energy as it passes between the moving blades. This is because the passage-way between the blades is such that it facilitates pressure drop, due to which, steam is accelerated through the passage by reaction. The result of this reaction force increases the propelling force given to the turbine shaft by the rotor. The cross-sectional area of the blades changes in a converging manner. The structure of nozzles in the reaction turbine differs from that of a simple impulse turbine. Rather, they resemble two-dimensional channels created in the space between each row's blades. The reaction turbine obtains its total decrease in pressure gradually and continuously by passing through a sequence of fixed and moving blades. The stationary blades operate as guide vanes to control the steam flow direction and increase the steam velocity. The Parson's turbine is a specific type of turbine that shows the features of an impulse-reaction turbine.

5.2 Difference between Impulse & Reaction Turbines

The key difference between impulse & reaction turbines are following:

- Flow of Steam

In an impulse turbine, steam is directed via a nozzle and impacts the turbine blades directly. While a reaction turbine guides the steam using fixed blades, causing it to flow across the turbine.

- Type of force on blade

In an impulse turbine, the rotation of the turbine is caused by an impulsive force. However, in a reaction turbine, the rotation is a result of the reaction force acting on the blades.

- Pressure of water

In an impulse turbine, the pressure of steam remains constant and equal to atmospheric pressure as it passes over the rotating blades. However, in a reaction turbine, the pressure of steadily falls as it flows over the blades.

- Pressure before striking the blade

In an impulse turbine, the pressure of steam is transformed into kinetic energy before striking the turbine blades. However, in a reaction turbine, there is no change in the steam pressure before it strikes the turbine blades.

- Flow rate

Impulse turbines are appropriate for relatively low steam flow rates, whereas reaction turbines are excellent for applications with higher steam flow rates.

- Casing

The presence of a turbine casing is necessary in a reaction turbine due to the significant difference in pressure between the intake and exit. Consequently, the shell effectively prevents the entry of air pressure.

- Blade profile

The blades of an impulse turbine have a symmetrical profile whereas blades of a response turbine are asymmetrical and have an aerofoil shape.

- Discharge

In an impulse turbine, the water is discharged straight from the turbine wheel into the tail race but water in a reaction turbine is first discharged into a draft tube and then into the tail race.

5.3 Simple De-Laval Turbine

A simple impulse turbine is often referred to as the De-Laval turbine, named after its inventor. The system has a single array of nozzles and a collection of rotating blades, as seen in Figure 5.4. The steam undergoes expansion from its beginning pressure to final pressure inside a single set of nozzles. As a result of a significant decrease in pressure inside the nozzles, the velocity of the steam within the nozzle increases. This high velocity steam from nozzle strikes the blades mounted on wheel. This high velocity soon decreases but the pressure remains constant during movement of steam over blades.

This turbine uses fixed nozzles to allow the steam to expand completely from boiler pressure to condenser pressure. As a result, the kinetic energy of the steam increases and its enthalpy decreases, producing shaft work. The steam moves on rotating blades at constant pressure but with a gradual decrease in velocity. The steam leaving the nozzles of this turbine has a uniform velocity of around 1050 m/s. A single row of blades subjected to this high-velocity steam jet could get the rotor speed reach as high as 30,000 rpm, which is too high for usage. Furthermore, there is a loss of energy called as carry-over loss or leaving loss since the steam is leaving the rotating blade at a high velocity. As a result, a single row of blades cannot efficiently use the steam jet's velocity. Therefore, it is essential to make certain modifications. The method is to use many sets of nozzles, blades, and rotors in sequence so that the turbine may gradually control the steam jet velocity or pressure.

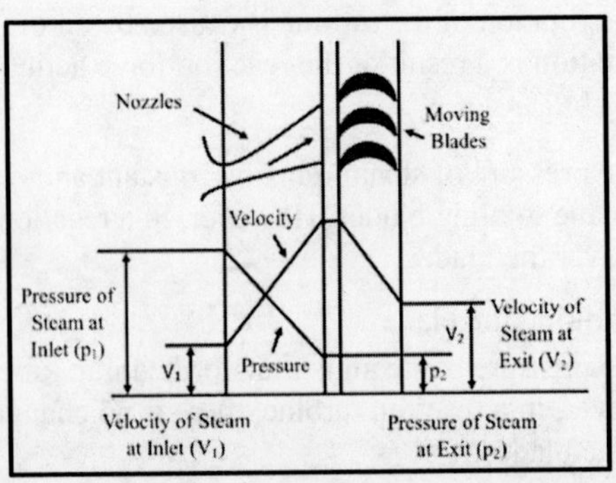

Fig. 5.4 Schematic of Simple De-Laval turbine

5.3.1 Velocity Diagram

When steam is supplied to an impulse turbine with a single wheel, it expands entirely in the nozzle and leaves at a high absolute velocity. The rotor faces towards the steam at an angle of α. Consequently, the absolute inlet velocity consists of two components: tangential and axial. The axial (vertical) component, also known as the axial or flow velocity, allows the steam to flow over the wheel, while the tangential component, also known as the whirl velocity (with direction of rotation), is primarily responsible for usable work. The velocity of the steam with respect to the blade at the entry point is known as the relative velocity, and it forms an angle θ with the blade as the blades moving at a velocity of u. In general, the following concepts are used in the vector diagram of velocity shown in figure 5.5:

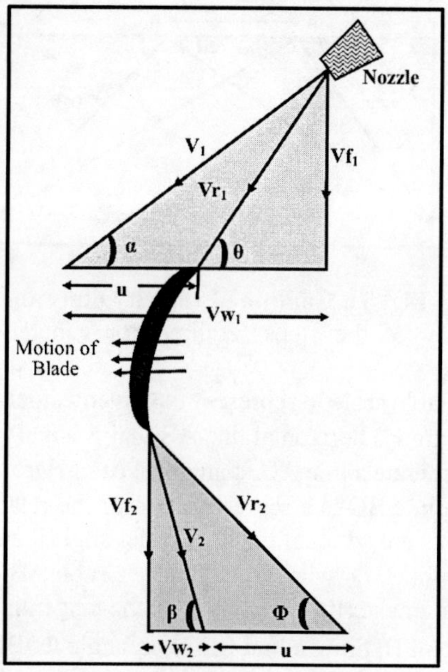

Fig. 5.5 Velocity diagram

The nomenclature for vector diagrams is as follows:

V_1, V_2= Absolute velocity of steam leaving the nozzle and blade respectively in m/s

α = Angle of steam jet in respect to plane of blade rotation

u = Linear velocity of blade in m/s

V_{r1}, V_{r2} = Relative velocity of steam at inlet and outlet in m/s

θ = Blade inlet angle

V_{w1}, V_{w2}= = Whirl velocity (component V_1, V_2 in the direction of rotation) in m/s

V_{f1}, V_{f2}= Axial or flow velocity of steam and Axial velocity at discharge in m/s

β = Steam jet outlet angle

Φ = Blade outlet angle

In order to calculate the relative velocity at the inlet, the vector difference between the absolute velocity V_1 and the blade velocity u should equal to the relative velocity V_{r1}. When compared to the axis of rotation, the relative velocity V_{r1} forms an angle of θ. As a result of the fact that steam does not expand inside the moving blades, the relative velocity V_{r2} at the exit, in the absence of any frictional effects, will be equivalent to the relative velocity at the inlet. On the other hand, if there is friction, then the two relative velocities form a relationship as follows:

$$V_{r2} = kV_{r1}$$

Here k is blade velocity constant with values ranging from 0.7 to 0.9.

5.3.2 Combined Velocity Diagram

As we know that linear velocity of blade (u) is common in both vector diagrams as shown in earlier figure, the combined velocity diagram can be drawn with the help of u.

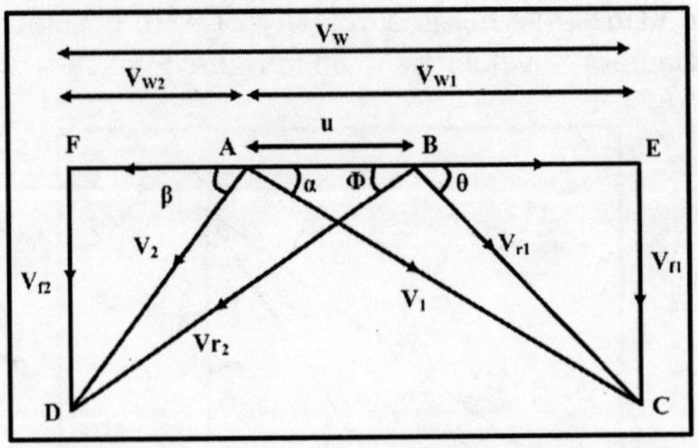

Fig. 5.6 Combined velocity diagram

Following is the summary of the steps required for generating the combined velocity diagram:

- Firstly, select a suitable scale to represent the given velocities on paper.
- The first step is to draw a horizontal line AB that is equal to the blade velocity u.
- Now, using AB, generate a line AC at an angle of α. Here AC = V_1.
- Join B and C. The line BC is a representation of the relative velocity at the entrance (V_{r1}). Measurements are taken of the blade inlet angle θ and the value is recorded.
- Beginning from point C, draw a perpendicular CE on AB extended. CE stands for the axial velocity at the intake (V_{f1}), while AE is the tangential velocity at the inlet.
- Beginning from point B, draw a line BD at an angle Φ. BD is equal to V_{r2}. Now, join A and D. The line AD stands for the absolute velocity at the outlet. Measurements are taken and the angle β is noted down.
- Beginning from point D, draw DF perpendicular to BA. If this is the case, then AF stands for the tangential velocity of steam at the outlet, and DF stands for the axial velocity at the outlet (V_{f2}).

➢ **Axial Thrust**

The difference in the velocities of flow (axial velocity) at entry and blade output results in axial thrust on the wheel. The wheel's axial thrust must be balanced, or a thrust bearing may be used to handle the load. The axial thrust can be calculated by following relation:

Axial thrust = mass flow rate × change in axial velocity

Axial thrust = $\dot{m} \times \left(V_{f1} - V_{f2}\right)$

➢ **Tangential Thrust**

Tangential thrust = Rate of change of momentum in the tangential direction

Tangential thrust = mass per second × change in velocity

Tangential thrust = $m \times (V_{w1} \pm V_{w2})$

- **Blade or Diagram Efficiency**

In De-Laval turbine, the total energy supplied to the blades is equal to the kinetic energy of steam entering the turbine. The blade efficiency is also known as diagram efficiency (η_b) and can be expressed as

$$\eta_b = \frac{Work\ done\ on\ blade}{Enegy\ supplied\ to\ the\ blade}$$

$$\eta_b = \frac{uV_w}{V_1^2/_2} \text{ or } \eta_b = \frac{2uV_w}{V_1^2}$$

Here V_w is the whirl velocity.

- **Stage Efficiency & Nozzle Efficiency**

The stage efficiency relates work done in the stage with the enthalpy decrease in the nozzle. It can be expressed as

$$\eta_{stage} = \frac{Work\ done\ on\ blade}{Enegy\ supplied\ per\ stage}$$

$$\eta_{stage} = \frac{uV_w}{\Delta h}$$

Here Δh is the specific enthalpy drop.
Also, Stage efficiency = Blade efficiency × Nozzle efficiency
∴ Nozzle efficiency = Stage efficiency/ Blade efficiency

$$\eta_{nozzle} = \frac{Stage\ Efficiency}{Blade\ Efficiency}$$

$$\eta_{nozzle} = \frac{\eta_{stage}}{\eta_{blade}}$$

$$\eta_{nozzle} = \frac{V_1^2}{2(h_i - h_e)}$$

Here h_i and h_e are the enthalpy at the inlet and exit respectively in J/kg.

5.4 Methods of reducing rotor speed

In the case of a simple impulse turbine, the steam undergoes expansion from the pressure in the boiler to the pressure in the condenser in a single step. Therefore, the rotor's speed becomes very high for practical applications. Steam turbines are compounded to achieve a practical rotor speed. Compounding is a process where many sets of rotors are connected to a single shaft in a sequential manner, allowing the steam pressure or jet velocity to be absorbed in stages as it passes over the rotor blades. The rotor speed of a steam turbine may be decreased using several ways of compounding.

(a) Velocity compounding

(b) Pressure compounding

(c) Pressure-velocity compounding.

➤ Velocity Compounding

The steam undergoes expansion through a fixed nozzle, transitioning from the initial pressure at the boiler or input to the lower pressure at the condenser. As the pressure decreases in the nozzle, the steam's kinetic energy rises as a result of its velocity increasing. A fraction of this accessible energy is absorbed by a series of moving blades. The steam, whose velocity has diminished while passing past the rotating blades, subsequently passes through the stationary second row of blades. The purpose of these stationary blades is to redirect the steam flow without changing its speed, directing it towards the next set of rotating blades. In this subsequent stage, work is performed on the blades, causing the steam to exit the turbine with a reduced velocity. Figure 5.7 (a) shows a cross-sectional view of a stage, illustrating the changes in pressure and velocity as steam flows through the nozzle, fixed blades, and moving blades.

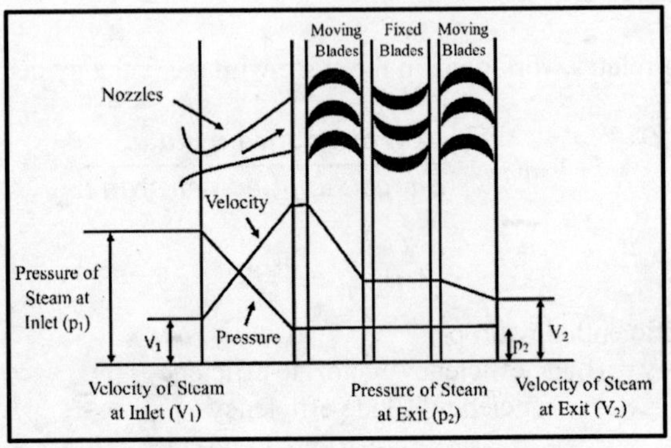

Fig. 5.7 (a) Velocity Compounding

➤ Pressure Compounding

The representation of the changes in pressure and velocity of steam may be seen in Figure 5.7 (b). The structure has a series of stationary nozzles that are integrated between the rotating blades' rings. The rotating blades are securely attached to the shaft. In this case, the complete pressure decrease occurs not in a single set of nozzles, but rather in many phases. Every stage comprises a collection of nozzles and a circular arrangement of rotating blades. The steam from the boiler flows through the first set of nozzles, undergoing partial expansion. Almost all of its velocity is dissipated when it traverses the first row of the mobile blades. It then proceeds to the second set of stationary nozzles and undergoes partial expansion once again. Subsequently, the steam is directed towards the second row of rotating blades, where its velocity is once again almost completely dissipated. This operation is iterated until the steam is discharged at the pressure of the condenser. By gradually decreasing the pressure, the speed of steam entering the rotating blades is significantly reduced. Therefore, the velocity of the rotor is decreased.

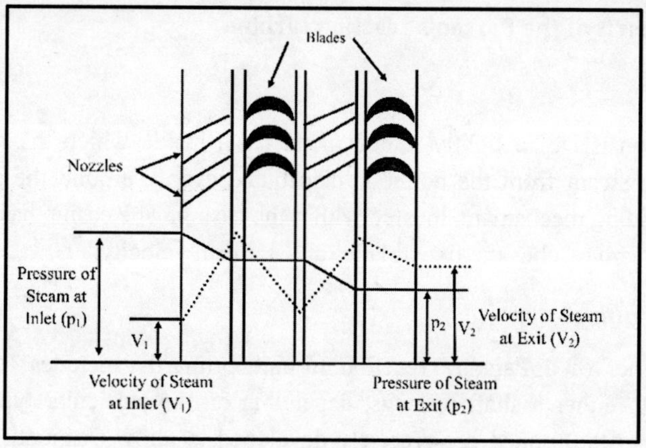

Fig. 5.7 (b) Pressure Compounding

➢ Pressure-Velocity Compounding

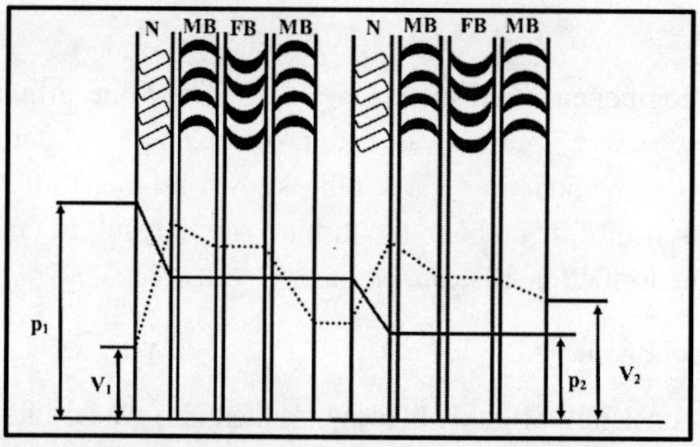

Fig. 5.7 (c) Pressure - Velocity Compounding

The pressure and velocity compounding of steam is shown in Figure 5.7 (c). This approach employs both pressure and velocity compounding techniques. The decrease in steam pressure occurs in two phases, and the resulting velocity is also combined in each step. The nozzles are immovably positioned at the start of each cycle, and the pressure stays constant throughout each stage.

5.5 Parson's Reaction Turbine

The Parson's reaction turbine, often referred to as a simple reaction turbine, operates by allowing pressurized steam to enter the turbine wheel and run over its blades. The steam, as it flows smoothly, forces the blades to move. However, the turbine runner is pushed by the reactive forces generated by steam jets. The downward motion of the blades is similar to the recoil of a gun. Technically, this turbine is based on the fundamental principle of Newton's third law of motion. The steam is introduced into the turbine with both high pressure and high velocity. As the steam passes through the nozzles, it undergoes expansion and increases in velocity. The high-velocity steam then reaches the rotor blades, inducing rotational motion.

5.5.1 Principle parts of the Parson's reaction turbine

- **Casing**

The apparatus consists of a tightly sealed metal case, inside which the high-pressure and high-temperature steam from the boiler is distributed evenly around the stationary blades, known as the guiding mechanism, located within the casing. The casing has been designed to ensure that the steam reaches the fixed blades at a uniform velocity.

- **Turbine runner**

The turbine runner of a Parson's reaction turbine primarily includes runner blades that securely attach to either a shaft or rings, depending on the particular type of turbine. The blades, attached to the runner, are precisely developed to allow a smooth entry and exit of steam without causing any abrupt disturbances. The turbine runner's surface is carefully polished to reduce frictional losses. Typically, the turbine runner is forged as a single unit. However, sometimes, it consists of individual steel plates that are welded together.

- **Draft tube**

A draft tube is a component used in turbines to improve the efficiency of the turbine by converting the kinetic energy of the steam into pressure energy. After crossing the runner, the steam proceeds into the condenser via a tube known as the draft tube. It is important to observe that the absence of this tube in the turbine would result in uncontrolled movement of steam, leading to the formation of steam eddies.

- **Guide mechanism**

It is a mechanism consisting of guide blades arranged in a wheel-like structure. The wheel is often attached to the casing, which is why these guide blades are sometimes referred to as fixed blades. The guide blades are precisely engineered to facilitate the smooth entry of steam into the runner, minimizing the possibility of shock. This is achieved by maintaining the velocity of the fluid entering the turbine runner tangential to the angle of the turbine blades. Also, it is necessary to ensure that the turbine receives the necessary amount of steam. This is accomplished by modifying the openings of the blades.

5.5.2 Velocity Diagrams of the Parson's reaction turbine

Steam, in the shape of a jet, enters the curved blade at point C after leaving the fixed blade. The jet should smoothly move over the inner surface and exit the blade at point D, as seen in Figure 5.8. Now, the velocity triangles at the entrance and exit points of the moving blade can be seen in Figure 5.8.

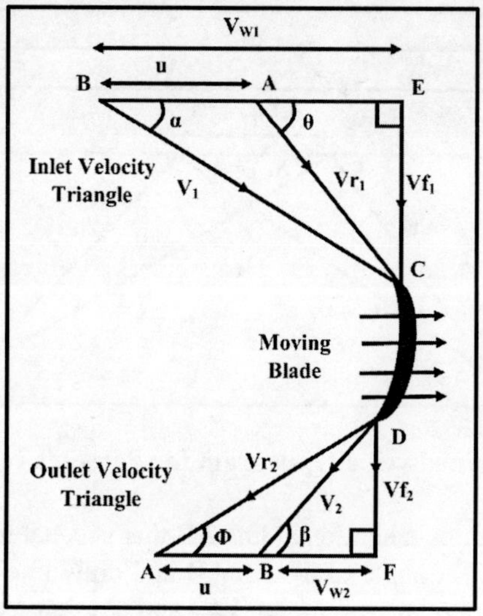

Fig. 5.8 Velocity diagram of Parson's reaction turbine

The nomenclature for vector diagrams is following:
V_1, V_2= Absolute velocity of steam leaving the nozzle and blade respectively in m/s
α = Angle of steam jet in respect to plane of blade rotation
u = Linear velocity of blade in m/s
V_{r1}, V_{r2} = Relative velocity of steam at inlet and outlet in m/s
θ = Blade inlet angle
V_{w1}, V_{w2}== Whirl velocity (component V_1, V_2 in the direction of rotation) in m/s
V_{f1}, V_{f2}= Axial or flow velocity of steam and Axial velocity at discharge in m/s
β = Steam jet outlet angle
Φ = Blade outlet angle

The jet of steam moves towards the blades along line BC having velocity V_1 at angle α along the direction of the blade motion. The axial component of V_1 performs no work on the blade is known is velocity of flow and denoted by V_{f1}. The key role of this component is to make steam flow through the blade and provide axial thrust on rotor. The linear velocity *u* of blade in m/s is represented by AB. The relative velocity of the steam is represented by V_{r1} w.r.t. blade and after passing over the blade, the relative velocity will become V_{r2}. The line AC represents the same. The absolute velocity of jet (V_2) as it leaves the blade is represented by DB inclined at an angle β with the direction of blade motion. The tangential component of V_2, (represented by BF) is known as velocity of whirl at exit (V_{w2}). The axial component of V_2,, (represented by DF) is known as velocity of flow at exit (V_{f2}).

5.5.3 Combined Velocity Diagrams of the Parson's reaction turbine

In reaction turbines, the steam keeps expanding as it moves over the rings of fixed and moving blades. It is achieved by using a tapered rotor with blade heights that become higher and higher. When the steam expands around the moving blade, it speeds up the relative velocity at the exit. The relative velocity V_{r2} is always higher than the relative velocity V_{r1} at the point of entry. Figure 5.9 shows the entry and exit velocity graphs with a shared blade velocity u.

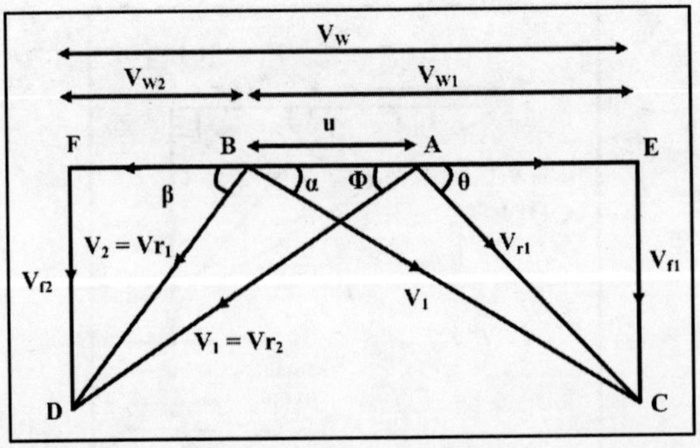

Fig. 5.9 Combined velocity diagram for Parson's reaction turbine

- The first step is to draw a horizontal line AB that is equal to the blade velocity u.
- Now, draw line BC at angle of α with AB and draw line AC with angle θ with EA. Here CA and CB are relative velocity (V_{r1}) and absolute velocity (V_1) at inlet.
- Now, draw line AD from A at angle Φ with AB and draw line BD at angle β with AB. Here DA and DB are relative velocity (V_{r2}) and absolute velocity (V_2) at exit.
- Now draw perpendicular from point C and D to meet at E and F.
- Here EB and CE are whirl velocity (V_{w1}) and velocity of flow (V_{f1}) at inlet and BF and DF are whirl velocity (V_{w2}) and velocity of flow (V_{f2}) at exit.

➢ Degree of Reaction

The degree of reaction in a reaction turbine is determined by calculating the ratio of the isentropic enthalpy drop in the moving blades to the isentropic enthalpy drop in the whole stage as shown in Figure 5.10.
The degree of reaction (R) may be determined using the following equation:

$$R = \frac{\text{enthalpy drop in the moving blades}}{\text{enthalpy drop in the whole stage}}$$

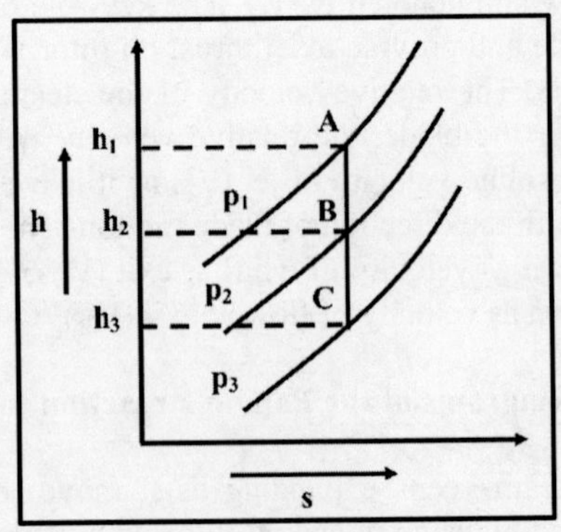

Fig. 5.10 h-s chart

$$R = \frac{h_2 - h_3}{h_1 - h_3}$$

The enthalpy drop in fixed blade (per kg. of steam) is given by

$$h_1 - h_2 = \frac{V_1^2 - V_2^2}{2000}$$

The enthalpy drop in moving blade (per kg. of steam) is given by

$$h_2 - h_3 = \frac{V_{r2}^2 - V_{r1}^2}{2000}$$

Now, total enthalpy drop will be

$$h_1 - h_3 = (h_1 - h_2) + (h_2 - h_3)$$

$$h_1 - h_2 = \frac{V_1^2 - V_2^2}{2000} + \frac{V_{r2}^2 - V_{r1}^2}{2000}$$

$$h_1 - h_3 = 2(h_2 - h_3) \text{ kJ/kg}$$

$$\text{As we know } R = \frac{h_2 - h_3}{h_1 - h_3}$$

$$\therefore R = \frac{h_2 - h_3}{2(h_2 - h_3)} = \frac{1}{2} = 0.5 \text{ or } 50\%$$

➢ **Blade or diagram efficiency**

The blade or diagram efficiency can be obtained as:

$$\eta_b = \frac{Work\ done\ in\ stage}{heat\ or\ energy\ supplied}; \quad \eta_b = \frac{w}{\Delta h}$$

$$\eta_b = \frac{u\,(V_{w1} + V_{w2})}{\Delta h}$$

As per velocity diagram, the work done will be; $w = u(V_{w1} + V_{w2})$
Here $V_{w1} = V_1 cos\alpha$ and $V_{w2} = V_{r2} cos\Phi - u$
So, $w = u(V_1 cos\alpha + V_{r2} cos\Phi - u)$
Also, $\alpha = \Phi$ and $V_{r2} = V_1$

$$\therefore w = u(V_1 cos\alpha + V_1 cos\alpha - u)$$

$$w = V_1^2\left(\frac{2uV_1}{V_1^2} cos\alpha - \frac{u^2}{V_1^2}\right)$$

$$w = V_1^2(2s\ cos\alpha - s^2)$$

$$(\because blade\ speed\ ratio, s = {}^{u}/_{V_1})$$

Δh is the sum of the kinetic energy supplied to the fixed and moving blades.

$\therefore$ the kinetic energy supplied to the fixed blades is ${}^{V_1^2}/_2$ and kinetic energy supplied to the moving blades is $\frac{V_{r2}^2 - V_{r1}^2}{2}$.

So, $\Delta h = \frac{V_1^2}{2} + \frac{V_{r2}^2 - V_{r1}^2}{2}$; $\Delta h = \frac{V_1^2}{2} + \frac{V_1^2 - V_{r1}^2}{2}$ $(\because V_{r2} = V_1)$ $\Delta h = V_1^2 - \frac{Vr_1^2}{2}$
In velocity Δ ABC, $Vr_1^2 = u^2 + V_1^2 - 2uV_1 cos\alpha$
Using above relation, Δh becomes

$$\Delta h = V_1^2 - \frac{u^2 + V_1^2 - 2uV_1 cos\alpha}{2}$$

$$\Delta h = \frac{V_1^2}{2} - \left\{1 - \left(\frac{u}{V_1}\right)^2 + 2\frac{u}{V_1}\cos\alpha\right\}$$

$$\Delta h = \frac{V_1^2}{2} - \left\{1 - (s)^2 + 2\frac{u}{V_1}\cos\alpha\right\}$$

Now, the blade or diagram efficiency will be

$$\eta_b = \frac{V_1^2(2s\cos\alpha - s^2)}{\frac{V_1^2}{2} - \left\{1 - (s)^2 + 2\frac{u}{V_1}\cos\alpha\right\}}$$

$$\eta_b = \frac{2(2s\cos\alpha - s^2)}{\{1 - (s)^2 + 2s\cos\alpha\}}$$

$$\eta_b = \frac{2(1 + 2s\cos\alpha - s^2) - 2}{1 + 2s\cos\alpha - s^2}$$

$$\eta_b = 2 - \frac{2}{1 + 2s\cos\alpha - s^2}$$

The blade or diagram efficiency will be maximum if value of $(1 + 2s\cos\alpha - s^2)$ becomes maximum. The maximum value can be obtained by differentiation.

$\therefore \frac{d}{ds}(1 + 2s\cos\alpha - s^2) = 0;\ 2s\cos\alpha - 2s = 0$ or $s = \cos\alpha$

Using the above value, the maximum blade efficiency will be

$$\eta_{bmax} = 2 - \frac{2}{1 + 2\cos^2\alpha - \cos^2\alpha}$$

$$\eta_{bmax} = \frac{2 + 2\cos^2\alpha - 2}{1 + \cos^2\alpha}$$

$$\eta_{bmax} = \frac{2\cos^2\alpha}{1 + \cos^2\alpha}$$

- **Blade height**

The blade height in a reaction turbine can be calculated by determining continuity equation. Through continuity equation, the mass flow rate of steam can be given as

$$\dot{m}_s = \frac{flow\ velocity \times area\ of\ flow}{specific\ volume\ of\ steam}$$

$$\dot{m}_s = \frac{V_{f1} \times 2\pi r h}{v_g}$$

Here specific volume of steam is v_g at steam pressure; area of flow is product of mean circumference of rotor and blade height.

Numerical 5.1 The Parson reaction turbine generates 75 kW per kg of steam when it operates at 400 rpm with a 50% reaction rate. The steam velocity is 1.4 times the blade velocity, and the blade exit angle is 20°. Find the blade's (a) velocity and (b) inlet angle.

Solution: Given $\alpha = \Phi = 20°$; $\dot{m}_s = 1$ kg/s; $V_1 = 1.4\,u$; N= 400 rpm; P = 75 kW; A= 0.5

Refer Fig. 5.8; $V_{w1} = V_1\cos\alpha$; $V_{w1} = V_1\cos 20° = 0.9397V_1$

$$V_{w1} = 0.9397 \times 1.4\,u = 1.31558\,u$$

Now, Power can be obtained as

$$P = \frac{\dot{m_s} \times u \times (V_{w1} + V_{w2})}{1000}$$

$$P = \frac{1 \times u \times (1.31558\, u + 0.31558\, u)}{1000}$$

$$u^2 = 45.979 \times 10^3$$

$$Blade\ Velocity\ (u) = 214.42 m/s$$

For blade angle, V_1= 1.4 u = 1.4×214.42

$$V_1 = 300.2\ m/s$$

$$V_{f1} = V_{f2} = V_1 sin\ \alpha$$

$$V_{w1} = V_1 cos\ \alpha;\ tan\theta = \frac{V_{f1}}{V_{w1} - u} = \frac{V_1 sin\ \alpha}{V_1 cos\ \alpha - u}$$

$$tan\theta = \frac{300.2\ sin20\,°}{300.2\ sin20\,° - 214.42} = 1.5171$$

$$\theta = 56.61\,°$$

Numerical 5.2 In a simple impulse turbine, the steam expands isentropically from 12 bar, 250°C, with an enthalpy of 2935 kJ/kg to 0.1 bar, with an enthalpy of 2584 kJ/kg. The blades are symmetrical and the nozzle makes a 20° angle with blade motion. Determine the blade velocity at 3600 rpm that provides maximum efficiency. Suppose that the steam has a negligible velocity as it enters the nozzle.

Solution: Given, p_0 = 12 bar; T_0 = 250°C; p_b = 0.1 bar; h_0= 2935 kJ/kg; h_1= 2584 kJ/kg; α= 20°; N = 3600 rpm; V_0= 0; Φ = θ

Now, change in specific enthalpy of steam is $\Delta h = h_0 - h_1 = 2935 - 2584 = 351 kJ/kg$

The steam jet velocity; V_1 = 44.72$\sqrt{\Delta h}$= 837.83 m/s

For maximum velocity; $\frac{Cos\alpha}{2} = {}^{u}/_{V_1};\ u = \frac{V_1 Cos\alpha}{2}$

$$u = \frac{837.83 \times cos20\,°}{2}$$

$$Blade\ velocity\ (u) = 393.65\ m/s$$

Numerical 5.3 Show that for a Parson's reaction turbine, the degree of reaction is 50%.
Solution: The degree of reaction can be defined as ratio of enthalpy drop in moving blade to the total enthalpy drop in stage.

$$\text{Degree of reaction } = \frac{\Delta h_m}{\Delta h_m + \Delta h_f}$$

Here $\Delta h_m = \frac{V_{r2}^2 - V_{r1}^2}{2}$ and $\Delta h_f = \frac{V_1^2 - V_2^2}{2}$

Using velocity diagram as shown in Fig. 5.8; θ = β and Φ = α

The inlet and outlet velocity diagrams are symmetrical with common blade velocity u.
Therefore; $V_1 = V_{r1}$ and $V_2 = V_{r2}$

$$\text{Now; } \Delta h_f = \frac{V_{r2}^2 - V_{r1}^2}{2}$$

Total enthalpy drop can be written as $\Delta h_m + \Delta h_f = \frac{V_{r2}^2 - V_{r1}^2}{2} + \frac{V_{r2}^2 - V_{r1}^2}{2}$

$$\Delta h_m + \Delta h_f = 2\Delta h_m$$

$$\text{Now, Degree of reaction} = \frac{\Delta h_m}{2\Delta h_m} = \frac{1}{2} \text{ or } 50\%$$

Numerical 5.4 The steam, which is at a pressure of 5 bar and a temperature of 200° C, passes through nozzles. The steam is then fed into an impulse turbine at a flow rate of 30 kg/minute. The steam gets released into a condenser operating at a pressure of 0.2 bar. The velocity of the blade is 300 m/s. The nozzles are oriented at an inclination of 25° relative to the direction of motion of the blades, and the exit blade angle is 35°. Calculate the theoretical power generated by the turbine, assuming no friction is present.

Solution: Given: p_1= 5 bar; p_2= 0.2 bar; T_1= 200° C; u = 300 m/s; α = 25°; Φ = 35°; m = 30 kg/min or 0.5 kg/s

Using given values, the value of h_1 and h_2 can be calculate through Mollier diagram. Using h_1 and h_2; the heat drop can be calculated as

$$\Delta h = h_1 - h_2$$

$$\Delta h = 2850 - 2350$$

$$\Delta h = 510\ kJ/kg$$

Now, the steam velocity at inlet of blade will be, $V_1 = 44.72\sqrt{\Delta h}$

$$V_1 = 44.72\sqrt{510}$$

$$V_1 = 1010\ m/s$$

To obtain the theoretical power generated by the turbine, we need to calculate whirl velocity from the combined velocity diagram. The combined velocity diagram for present data can be drawn as:

a) The first step is to draw a horizontal line AB that is equal to the blade velocity (u) i.e. 300 m/s.
b) Now draw inlet velocity triangle ABC with base AB having angle α = 25° and $V_1 = 1010\ m/s$. Now using velocity triangle, V_{r1} can be calculated on scale as 850 m/s.
c) Now draw outlet velocity triangle ABD with base AB having angle Φ = 35°. Now using scale, $V_{r1} = V_{r2}$ i.e. 850 m/s.
d) Now draw perpendicular from point C and D to meet AB at E and F.

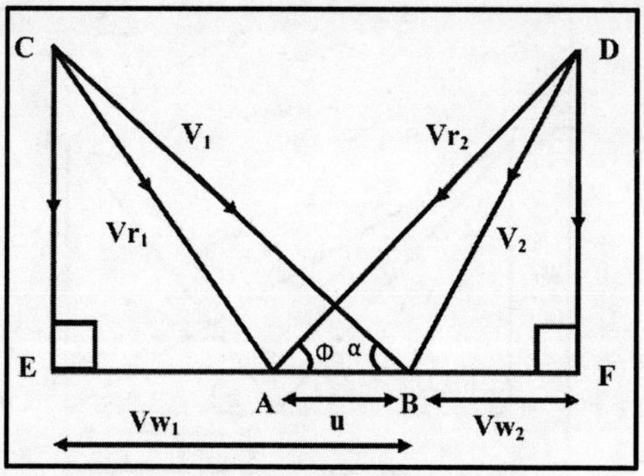

Using velocity triangle diagram, we can measure EF as

$$EF = V_{w1} + V_{w2} = 1400\ m/s$$

The theoretical power generated by the turbine will be

$$P = m(V_{w1} + V_{w2})\ u$$

$$P = 0.5 \times 1400 \times 300 = 210000\ W$$

$$P = 210\ kW$$

Numerical 5.5

Stream rate of an impulse turbine fitted with a single row of movable blades is 2 kg/s. The turbine generates a power output of 130 kW, with the blade velocity of 175 m/s. The steam flows from a nozzle at a speed of 400 m/s, while the blades have a velocity coefficient of 0.9. Determine the angle of the nozzle, as well as the angles of the blade at the entrance and exit, assuming that the steam flows axially after passing over the blades.

Solution: Given: P= 130 kW or 130×10³ W; u = 175 m/s; V_1 = 400 m/s; m = 2 kg/s; K = 0.9

We know Power; $P = m(V_{w1} + V_{w2})\ u$

$130 \times 10^3 = 2\ (V_{w1} + V_{w2})\ 175$ or $130 \times 10^3 = 2\ (V_w)\ 175$

$(\because (V_{w1} + V_{w2}) = V_w)$

$$V_w = \frac{130 \times 10^3}{350} = 371.4\ m/s$$

Based on the data given, we figure out that it is insufficient to draw up the inlet velocity triangle using the standard approach. Therefore, we must employ the geometry of the axial discharge. Now design the combined velocity triangle as:

a) The first step is to draw a horizontal line AB that is equal to the blade velocity (u) i.e. 175 m/s and EB equal to 371.4 m/s representing whirl velocity.
b) Now draw inlet velocity triangle EBC with base and $V_1 = 400\ m/s$. Join CA representing relative velocity V_{r1} which can be calculated on scale as 240 m/s.
c) Now draw outlet velocity triangle ABD with base AB having angle β = 90°. Now using scale, $V_{r1} = 0.9V_{r2}$ i.e. 216 m/s.

Using velocity triangle diagram, we can measure α, β and θ as 19°, 33° and 36° respectively.

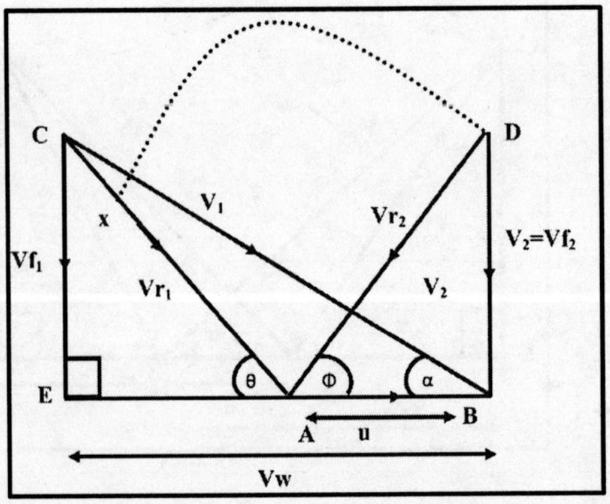

Numerical 5.6

In a reaction turbine, the mean diameter of the blade ring is 1 meter at a certain stage. The turbine operates at a speed of 3000 revolutions per minute. The blades are specifically designed to achieve a 50% degree of reaction, with exit angles of 30° and input angles of 50°. The turbine is provided with a steam flow rate of 10000 kg/h and operates with a stage efficiency of 85%. Calculate the following:
(a) The power output of the stage
(b) The specific enthalpy decrease in kJ/kg

Solution: Given: Degree of reaction = 50% or 0.5, Mean diameter = 1 m; N = 3000 rpm, $m =$ 10000 kg/h or 2.778 kg/s; η_{stage}= 0.85; α=Φ = 30°; θ=β= 50°

The blade velocity can be obntained as $u = \frac{\pi DN}{60} = \frac{3.14 \times 1 \times 3000}{60} = 157.08\ m/s$

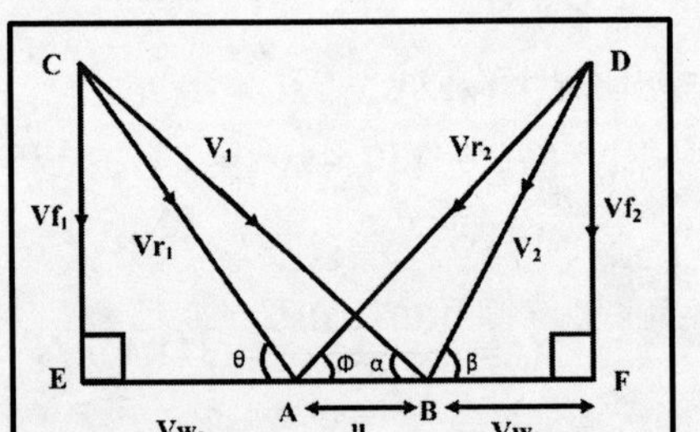

Using velocity diagram,

V_1= 351.8 m/s; V_{w1}= 304.7 m/s; V_{f1}= 229.6 m/s; V_{w2}= 157.62 m/s

Total whirl velocity (V_w)= $V_{w1} + V_{w2}$

$$V_w = 462.32\ m/s$$

$$\text{Now, } P = m(V_{w1} + V_{w2})\ u$$

$$P = 2.778 \times 157.08 \times 462.32 = 201741.2\ W\ Or\ 201.71\ kW$$

$$\eta_{stage} = \frac{uV_w}{\Delta h} \; Or \; \Delta h = \frac{uV_w}{\eta_{stage}} = \frac{157.08 \times 462.32}{0.85} = 85.43 \; kJ/kg$$

Numerical 5.7 The Parson reaction turbine has receiving tips with an angle of 35° and discharging tips with an angle of 20°. The velocity of the blade is 100 meters per second. Determine the tangential force, power generated, and axial thrust of the turbine given a steam consumption rate of 1 kg/min.

Solution: Degree of reaction = 50% or 0.5; u = 100 m/s, m = 1 kg/min or 1/60 kg/s; α=Φ = 20°; θ=β= 35°

The velocity diagram can be drawn as

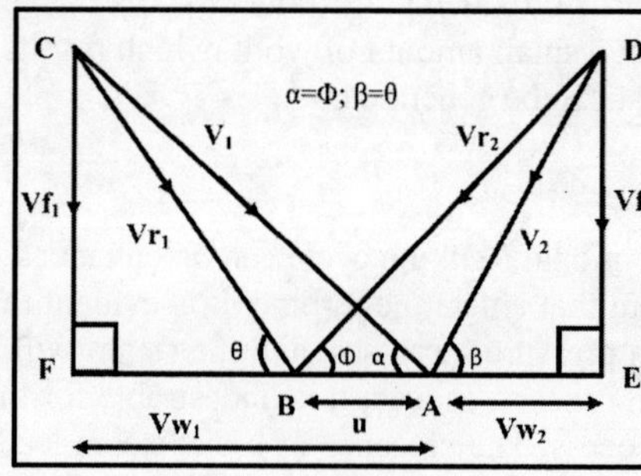

a) Draw a horizontal line AB that is equal to the blade velocity (u) i.e. 100 m/s.
b) Through A, draw line AC at angle of α=20°.
c) Through A, draw line AD at angle of β= 35°.
d) On the scale, measure AC equal to V_1= 220 m/s and BC as V_{r1}= 130 m/s.
e) Through point C, draw vertical line CF as inlet flow velocity and measure AF as V_{w1}= 130 m/s.
f) Since V_{r2}= V_1 and α=Φ; draw a line BD at an angle of 20°.
g) Join point A to point D representing as absolute velocity of steam at exit as V_2= 130 m/s.
h) Draw a line DE through D to represent flow velocity V_{f2} at exit.
i) Now, measure length of EA as V_{w2}= 107 m/s.

The tangential force can be calculated as = $m(V_{w1} + V_{w2})$

$$= \frac{1}{60}(207 + 107)$$

$$\text{Tangential force} = 5.233 \; N$$

The power generated will be $P = m(V_{w1} + V_{w2})\, u$

$$P = \frac{1}{60} \times 100 \times (207 + 107) = 524.4 \; W \; or \; 0.5244 \; kW$$

The axial thrust will be,
$$F_a = m\,(V_{f2} - V_{f1})$$

$$F_a = m(0) \quad (\because V_{f1} = V_{f2})$$

$$F_a = 0$$

5.6 Bleeding

Bleeding is the process of removing steam from the turbine at certain intervals throughout its expansion to utilize that steam to heat the boiler's feed water. When this process is carried out, a small quantity of steam is drained from the turbine from specific sections of the turbine. This steam is then cycled along the feed water line that connects the hot well to the boiler. However, since the water is relatively cold, the steam is able to condense, and the heat that is lost as a result of the steam is transferred to the feed water. The condensed steam eventually makes its way to the hot well.

The effects of bleeding include the following:

- It improves the turbine's thermodynamic efficiency while it is in operation.
- A supply of hot water is provided to the boiler.
- The turbine loses just a small amount of work, which results in a decrease in the amount of power that can be generated.

5.7 Re-heating

The efficiency of the conventional Rankine cycle can be enhanced by increasing the pressure and temperature of the steam that enters the turbine. It is evident that the expansion ratio will increase as the initial steam pressure increases, and the steam will become quite moist at the conclusion of the expansion process. Indeed, it is undesirable for the steam to get saturated at the end of the expansion process. The presence of moist steam leads to erosion of the turbine blades and results in higher internal losses. This will eventually reduce the efficiency of the turbine blades. To resolve the aforementioned challenge, the steam may be reheated. In this process, the steam is extracted from the turbine when it gets saturated. Subsequently, the steam is subjected to steady pressure and heated by the flue gases until it reaches the superheated condition once again. It is thereafter sent back to the subsequent phase of the turbine. Figure 5.11 shows the schematic representation of the reheating.

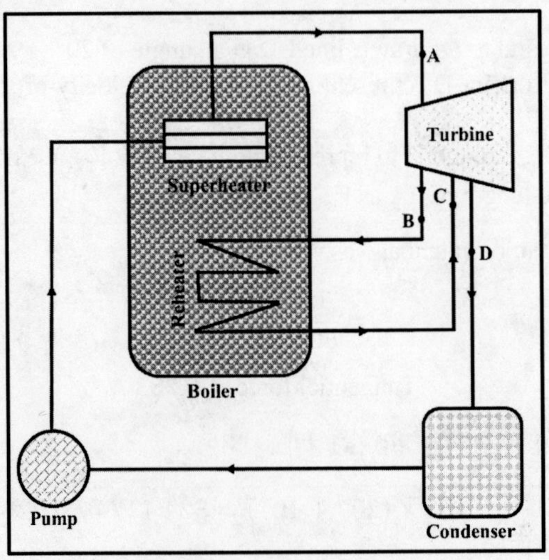

Fig. 5.11 Reheating of steam

Practically, the addition of reheat increases the efficiency of the cycle by around 5%. A second reheat will result in much reduced benefits, considering the excessive initial costs associated with implementing a two-stage reheat system. Therefore, the use of two-stage reheat is only feasible in circumstances of very high initial throttle conditions. The cost of reheat equipment, which includes a boiler, pipes, and controls, may be 5% to 10% more than that of conventional boilers. This extra expense is only justified if the increase in thermal efficiency is significant enough to provide a return on this investment.

The process of reheating steam in a turbine has the following benefits:
1. It enhances the output of the turbine.
2. It enhances the turbine's efficiency.
3. The increase in the dryness percentage of steam at the exhaust minimizes blade erosion.
4. The condenser of the turbine requires less water owing to a decrease in specific steam consumption.

5.8 Reheat Factor

The effect of friction on the flow of steam as it passes over the blades is to reduce the relative velocity of the steam. The energy that is lost as a result of friction is turned into heat, which in turn causes an increase in entropy and, as a result, a small decrease in the amount of heat is observed. According to the Mollier diagram, as shown in Figure 5.12, the effect of friction on the adiabatic flow of steam is effectively demonstrated. In this case, let A_1 stand for the initial phase of the steam. During this first phase, the steam undergoes adiabatic expansion, moving from pressure P_1 to pressure P_2. The vertical line A_1B_1 is used to show this point. To represent the heat that is produced as a result of blade friction, mark off the point C_1 such that B_1C_1 represents the value of heat generated. Using C_1 as a starting point, draw a horizontal line that will meet the back pressure (P_2) line at A_2. The point A_2 then depicts the final condition of the steam as it exits the first phase of the process. The second phase then consists of a repetition of the same method. At the conclusion of the second phase, the point A_3 represents the final condition of the steam, and the same concept applies to the other stages as well. The line A_1D is a representation of the adiabatic expansion of steam across all of the phases, if the friction is neglected.

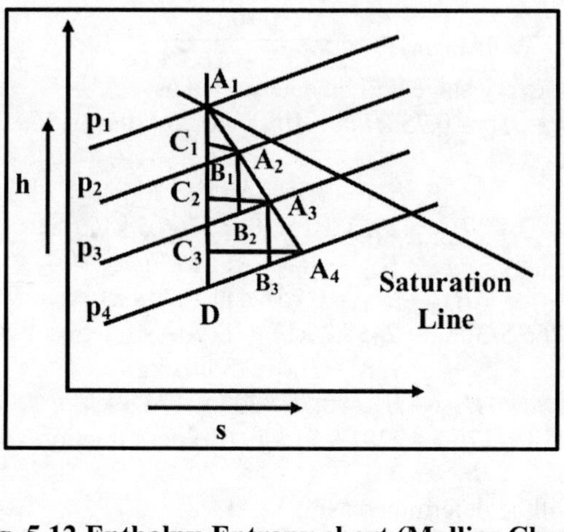

Fig. 5.12 Enthalpy-Entropy chart (Mollier Chart)

Now, the reheat factor can be calculated by obtaining the ratio of the cumulative heat drop to the isentropic heat drop. The sum of increased heat drops is known as cumulative heat drop.

$$\text{Reheat Factor; } F = \frac{\text{cumulative heat drop}}{\text{isentropic heat drop}}$$

$$F = \frac{A_1B_1 + A_2B_2 + A_3B_3}{A_1D}$$

(Note: Reheat factor is always greater than 1)

Numerical 5.8 The three stage turbine is fed with steam at a pressure of 15 bar and a temperature of 250°C. The steam is then exhausted at a pressure of 0.07 bar. Assuming that the efficiency of every stage is 75%, and the pressure at the intermediate stages is 5 bar and 0.4 bar, respectively. Calculate the Rankine efficiency, determine the steam condition at the end of every stage, the efficiency ratio, and reheat factor.

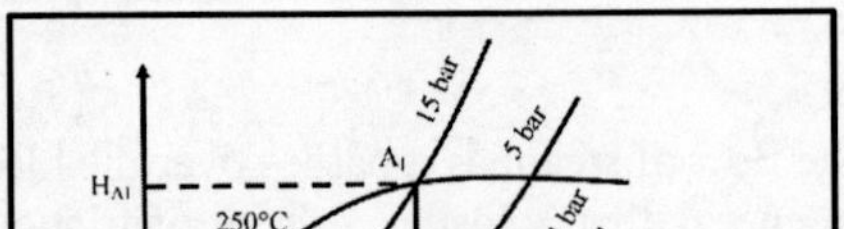

Solution:
Using Mollier diagram, $H_{A1} - H_D$= 840 kJ/kg; H_{A1} = 2923.3 kJ/kg; h_D= 163.4 kJ/kg;
$H_{A1} - H_{a1}$= 209 kJ/kg
The Rankine efficiency can be calculated as

$$\eta_{Rankine} = \frac{H_{A1} - H_D}{H_{A1} - h_D}$$

$$\eta_{Rankine} = \frac{840}{2923.3 - 163.4} = 30.4\%$$

The steam condition at every stage will be determined as

$$H_{A1} - H_{A2} = 0.75\,(H_{A1} - H_{B1}) = 0.75\times 209 = 156.8 \text{ kJ/kg}$$

$$H_{A1} - H_{A2} = 156.8 \text{ kJ/kg}$$

$$H_{A2} = 2923.3\text{-}156.8 = 2766.5 \text{ kJ/kg}$$

$$H_{A2} - H_{B2} = 411 \text{ kJ/kg}$$

$$H_{A2} - H_{A3} = 0.75\times 411 = 308 \text{ kJ/kg}$$

$$H_{A3} = 2766.5\text{-}308.3 = 2458.2 \text{ kJ/kg (dryness fraction } x_3 = 0.923)$$

$$H_{A3} - H_{B3} = 235 \text{ kJ/kg}$$

$$H_{A3} - H_{A4} = 0.75\times 235 = 176.3 \text{ kJ/kg}$$

$$H_{A4} = 2458.2\text{-}176.3 = 2281.9 \text{ kJ/kg (dryness fraction } x_4 = 0.879)$$

The efficiency ratio will be determined as

$$\eta_{ratio} = \frac{H_{A1} - H_D}{H_{A1} - h_D}$$

$$\eta_{ratio} = \frac{2923.3-2281.9}{840} = 0.7636 \text{ or } 76.36\%$$

Reheat Factor;

$$F = \frac{A_1B_1 + A_2B_2 + A_3B_3}{A_1D}$$

$$F = \frac{(H_{A1} - H_{B1}) + (H_{A2} - H_{B2}) + (H_{A3} - H_{B3})}{H_{A1} - H_D}$$

$$F = \frac{209 + 411 + 235}{840}$$

$$\text{Reheat Factor (F)} = 1.018$$

5.9 Governing of Steam Turbines

Steam turbine governing is the process of observing and controlling the steam flow rate into the turbine in order to keep its rotational speed constant. The valves are placed between the boiler and the turbine to monitor and regulate the flow rate of steam. The various forms of controlling procedures are being utilized depending on the specific mechanism used to manage the flow rate of steam. The main techniques for governance are outlined below.

- **Throttle Governing**

Throttle governing is extensively used, especially on small turbines, due to its lower initial cost and simple mechanism. The purpose of throttle governing is to maintain speed by reducing the steam whenever the load falls below the economic or design load, and vice versa. It works by reducing the steam pressure at the turbine inlet, which leads to a decrease in the amount of available energy. This technique includes the flow of steam along a restricted pathway, resulting in a decrease in pressure across governing valve. A partially opened steam control valve is used to regulate the flow rate. The decrease in pressure initiates a throttling phase where the enthalpy of steam remains constant.

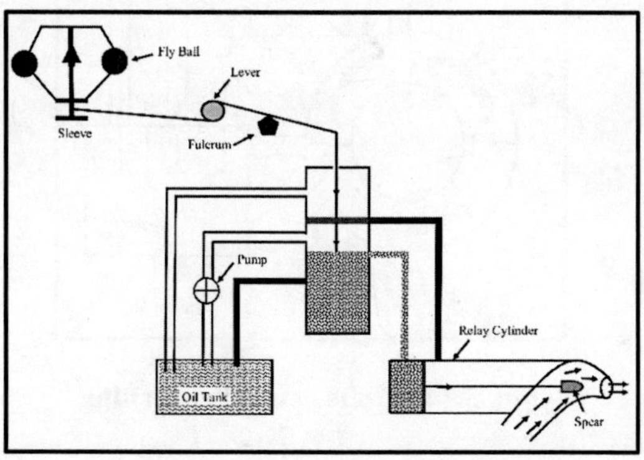

Fig. 5.13 Throttle Governing

The rotation speed of the governor ball increases with increase in speed of the engine. The resultant increase in centrifugal force propels them outward, opposing the force of gravity and the control spring. As a result, the inner ends of the arms of the governor balls apply pressure on the valve spindle, causing the valve to partially closed. This decreases the steam pressure which is exerted on the piston. As a result of the throttling at the valve, the power generated is decreased and the engine speed returns to its usual value. Similarly, a reduction in speed will result in a drop in weight and a rise of the valve spindle. The pressure exerted on the piston subsequently rises, resulting in an increase in the speed beyond its standard value.

➢ **Bye-pass Governing**

Bypass governing of a steam turbine involves the installation of a separate bypass line for the steam. This is particularly used when the turbine is operating under an excessive load. The bypass line is designed to transfer steam from the first stage nozzle box to the next stage where the work output is increased. This steam bypass is automatically controlled by the valve lift, which is regulated by the speed of the governor for all loads within its operational range. The bypass valve is opened to allow the discharge of new steam into the later stage of the turbine. In the final stages of production, there is an increase in work output but a decrease in efficiency due to the throttle effect.

➢ **Nozzle Control Governing**

In nozzle controlled governing, the nozzles are organized into sets consisting of two, three, or more groups. Each set of nozzles is regulated by an individual valve. When the load on the turbine decreases, it is possible to deactivate a certain number of nozzles that are needed. The nozzle control mostly impacts the first phase of the turbine. It has no effect on other phases. In nozzle control governing, the regulation of steam pressure is replaced by the regulation of the mass flow rate of steam. One benefit is that it utilizes steam at its maximum boiler pressure and reduces the rate of steam consumption while operating at lower loads. Figure 5.14 depicts a diagrammatic layout of nozzle control controlling. The system has three sets of nozzles labelled as N1, N2, and N3, with 5, 4, and 12 nozzles correspondingly. Valves V1, V2, and V3 regulate the activation and deactivation of these sets of nozzles.

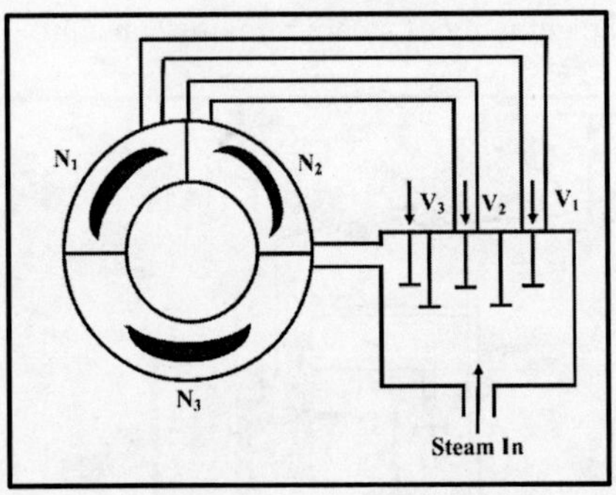

Fig. 5.14 Nozzle control governing

UNIT SUMMARY

Based on the contents covered in this unit you should remember that:

- ❖ There are three kinds of steam turbines: impulse turbines, impulse reaction turbines, and reaction turbines.
- ❖ The methods by which steam expands as it passes through impulse and reaction turbines are the primary differences between them.
- ❖ By incorporating numerous stages, steam turbine compounding reduces the rotational speed of an impulse turbine to practical speeds.
- ❖ Impulse steam turbines can be compounded by either velocity compounding, pressure compounding, or a combination of both.
- ❖ A difference in flow velocities at the entry and exit causes the axial thrust on the wheel.
- ❖ The degree of reaction in a reaction turbine stage is determined by dividing the heat drop along the rotating blades by the total heat drop in the stage.
- ❖ Whirl velocity is the tangential component of velocity, which is responsible for doing essential work.
- ❖ Flow velocity is the axial velocity component that enables the steam to flow over the wheel.
- ❖ Relative velocity can be defined as steam velocity relative to blade speed.
- ❖ The ratio of the work done on the blade to the energy provided at each stage is called stage efficiency.
- ❖ The ratio of the work done on the blade to the energy supplied to the blade is called blade efficiency.
- ❖ Steam turbine governing is the process of observing and controlling the steam flow rate into the turbine in order to keep its rotational speed constant.
- ❖ Reheat factor can be calculated by obtaining the ratio of the cumulative heat drop to the isentropic heat drop.

Exercises

Multiple Choice Questions

1. Which type of turbine is more suitable for high-speed applications?

   A) Impulse turbine

   B) Reaction turbine

   C) Both are equally suitable

   D) Neither, both are unsuitable for high-speed applications

2. Which turbine type generally requires a higher number of stages to achieve the desired pressure drop?

   A) Reaction turbine

   B) Impulse turbine

   C) Both require the same number of stages

   D) Neither requires multiple stages

3. In a Simple De Laval turbine, where does the steam undergo its primary expansion?
   A) In the turbine blades
   B) In the nozzles before reaching the blades
   C) In the stator blades
   D) In the condenser
4. What is a primary characteristic of the Simple De Laval turbine compared to other steam turbines?
   A) It uses multiple stages to achieve high pressure and temperature drops.
   B). It combines impulse and reaction principles in a single turbine stage.
   C) It achieves high efficiency by using a single stage with high-velocity steam
   D) It relies on gradual energy conversion through a series of expansion stages.
5. What is one common method used to reduce the speed of a turbine rotor?
   A) Increasing the steam pressure
   B) Using a gear reduction system
   C) Increasing the rotor diameter
   D) Increasing the steam temperature
6. Which of the following is NOT a method to reduce rotor speed in a steam turbine?
   A) Using a mechanical governor
   B) Employing a speed reducer gear
   C) Increasing the steam flow rate
   D) Adjusting the nozzle control to alter steam flow
7. Which component of the Parsons' Reaction Turbine is responsible for converting the steam's pressure into velocity?
   A) The turbine blades
   B) The nozzle
   C) The shaft
   D) The governor
8. What kind of steam flow arrangement is commonly used in Parsons' Reaction Turbines?
   A) Parallel flow
   B) Counter flow
   C) Axial flow
   D) Radial flow

9. **What is the main purpose of a reheating factor in the context of a steam turbine cycle?**

   A) To quantify the amount of heat added to the steam during reheating.

   B) To measure the efficiency of the bleeding process.

   C) To determine the cooling requirements of the steam turbine system.

   D) To calculate the pressure drop across the turbine stages.

10. **What does "bleeding" refer to in a steam turbine system?**

    A) The process of extracting steam from the turbine for use in heating or other purposes.

    B) The process of injecting additional steam into the turbine to increase power output.

    C) The technique of cooling the steam before it enters the turbine.

    D) The process of removing excess water from the steam.

Answer Key:

1 A	2 A	3 B	4 C	5 B
6 C	7 B	8 C	9 A	10 A

<u>Short and Long Answer Type Questions</u>

- What is the difference between an impulse turbine and a reaction turbine?
- What is the primary function of a steam turbine?
- Why do steam turbines use superheating?
- Discuss the differences between impulse and reaction turbines, including their design and operational characteristics.
- Explain the design and operational principles of Parsons' reaction turbine. What distinguishes it from other steam turbines and are its benefits?
- Describe the construction and significance of velocity diagrams in Parsons' Reaction Turbine. How do these diagrams help in analysing the turbine's performance?
- Derive the equations for the work done by the steam on the rotor blades using velocity triangles and explain their significance in the context of steam turbine operation.
- Discuss the causes of axial thrust in steam turbines and how turbine design manages it.
- Define and explain tangential thrust in steam turbines, including its impact on turbine performance and design considerations.
- In a steam turbine, discuss blade efficiency, its definition, and factors that affect it.

Know More

- ✓ Steam turbines are classified into two groups: high pressure and low pressure stages, in which high pressure is designed to manage steam at higher temperatures and pressures, whereas low pressure stages manage that steam, which undergoes expansion and cooling.
- ✓ The advanced modern steam turbines use the latest technology-based blades, which are more aerodynamic. These blades are designed with complex shapes and variable angles for optimized efficiency.
- ✓ Modern turbines use reheating processes to improve thermal efficiency. These turbines expand the steam at high pressure, reheat it, and then expand it at low pressure.
- ✓ The efficiency of steam turbines depends not only on the blade design, but also on the path that the steam follows through the turbine. Engineers use modern techniques to reduce the effect of friction, turbulence, and shock waves.
- ✓ Steam turbines often undergo several stages of development to enhance performance and efficiency. Every stage includes a set of rotor and stator blades. The staged design enables a steady expansion of steam, maximizing the efficiency of converting thermal energy into mechanical work.
- ✓ Steam turbines have dynamic loading conditions that could result in mechanical fatigue. Advanced analysis procedures, including fatigue testing and stress analysis, are used to anticipate and tackle any problems, ensuring long-term reliability and security.

References and Suggested Readings

- Thermal Engineering by B. R. Sarkar (McGraw-Hill Education (India) Pvt Limited).
- Thermal Engineering by R.K. Rajput (Laxmi Publications (P) Ltd).
- A Textbook of Thermal Engineering by R.S. Khurmi & J.K. Gupta (S. Chand & Company Ltd).
- Thermal Engineering by Mahesh. M. Rathore (McGraw-Hill Education (India) Pvt Limited).

Dynamic QR Code for Further Reading

1. Lectures of the NPTEL Course on "Impulse Steam Turbine: Velocity Diagrams, Work Transfer, Blade Efficiency by Dr P.K. Mondal, Department of Mechanical Engineering, Indian Institute of Technology Guwahati, India

2. Lectures of the NPTEL Course on "Reaction Steam Turbine: Velocity Diagrams, Work Transfer, Blade Efficiency by Dr P.K. Mondal, Department of Mechanical Engineering, Indian Institute of Technology Guwahati, India

3. Lecture on "Governing of steam turbines" from lecture 19-23 by Best Mechanical Engineering YouTube channel.

CO AND PO ATTAINMENT TABLE

Course outcomes (COs) for this course can be mapped with the programme outcomes (POs) after the completion of the course and a correlation can be made for the attainment of POs to analyse the gap. After proper analysis of the gap in the attainment of POs necessary measures can be taken to overcome the gaps.

Course Outcomes	Attainment of Programme Outcomes (1- Weak Correlation; 2- Medium correlation; 3- Strong Correlation)						
	PO-1	**PO-2**	**PO-3**	**PO-4**	**PO-5**	**PO-6**	**PO-7**
CO-1							
CO-2							
CO-3							
CO-4							
CO-5							

The data filled in the above table can be used for gap analysis.

Index